Fractals in Science

An Introductory Course

Pilot Edition

Center for Polymer Studies
Science and Mathematics Education Center
Boston University

Springer-Verlag New York

Software

Software is an essential part of student use of these materials. Software runs on the Macintosh line of computers and comes with a royalty-free license to copy for your own use. To receive software and updates on this text, send $25 US to the Center for Polymer Studies, 590 Commonwealth Avenue, Boston, MA 02215

This work was supported by grants from the National Science Foundation:
Applications of Advanced Technology (MDR 8955040, MDR 9112301)
Research in Teaching and Learning (MDR 9150076)
Teacher Preparation and Enhancement (ESI 9353500)
Instructional Materials Development (ESI 9353900)

Additional support was provided by:
Apple Computer Corporation
International Business Machines

Opinions expressed are those of the authors
and not necessarily those of the funders.

ISBN-13: 978-0-387-94361-9   e-ISBN-13: 978-1-4615-7012-7
DOI: 10.1007/978-1-4615-7012-7

Authors and Contributors

H. Eugene Stanley, Edwin F. Taylor, and Paul A. Trunfio
Coordinating Editors

Chapter 1. From Coin Flipping to Motion of Molecules
Chapter 2. Fractals in Nature
Chapter 3. Growth Patterns in Nature: Percolation
Richard H. Audet, Rama Bansil, Kenneth Brecher, Sergey V. Buldyrev, Mary E. Caggiano,
Frank Caserta, Galina Dobrynina, Melissa J. Erickson, Peter Garik, Sharon C. Glotzer, Mark F. Gyure,
Shlomo Havlin, Paul A. Hickman, Greg Huber, Charles L. Hurwitz, Joseph E. Jordan, Jyotsana Lal,
Hernan Larralde, Dennis E McCowan, C. K. Peng, Peter H. Poole, Srikanth Sastry, Stefan Schwarzer,
Francesco Sciortino, Robin Selinger, Mary H. Shann, Linda S. Shore, H. Eugene Stanley, Dietrich Stauffer,
Edwin F. Taylor, Paul A. Trunfio

Laboratory experiments developed and documented under the supervision of
Peter Garik, Kenneth Brecher, Mary E. Caggiano, assisted by Boston University students:
Maryellen Abraham, Melanie Cardoza, Weng Ki Ching, Suk Choi, Melissa C. Ferris, Paula Giardini,
Matthew Gillespie, Anthony Hill, Reshma Kewalramani, Vasudha Kuruganti, Chae Lee, Suzee Lee,
David Maas, Jennifer Mathieu, Daniel Michael, Julie Monagle, Mary Morrissey, Marilyn Negron,
Hong Noe, Laryssa Pranis, Y. Theresa Silta, John Taraszka, Liana Vesga, Renate Yang

Chapter 4. Literature, DNA, and Fractals
S. Martina Ossadnik, Sergey V. Buldyrev, Shlomo Havlin, Rosario N. Mantegna, Edwin F. Taylor

Chapter 5. Spin Glasses and Neural Networks
Sava Milosevic and Sona Prakash

Chapter 6. Lightning and Soap Films
Sava Milosevic and Marek Wolf

Chapter 7. Rough Surfaces
David Futer, Albert-Laszlo Barabasi, Sergey V. Buldyrev, Shlomo Havlin, Hernan Makse

Programmers
Nicholas Andrews, Leo Braginsky, Sergey V. Buldyrev, James Blandy, Frank Caserta, Dimitri Chernyak,
Anthony Collins, Roman Dultzin, Edmund Feingold, Morris E. Matsa, Tewolde H. Mekonen,
Boris Ostrovsky, Deborah Pearlman, Thanh Pham, Angela Ranieri, Airlee Satler, Michael Sherman,
Becky Steinberg, Peite Su, Edwin F. Taylor, Tom Udale, Michelle M. Ukleja, B. G. Volbright,
Thomas Weisbach, Inna Zarnitsky, Ruslana Zitserman

Kenneth Brecher and Gerald L. Abegg
Directors, Science and Mathematics Education Center

Contents

Contents

Contents

3. Growth Patterns in Nature: Percolation

Contents

Contents

Contents

Contents

Contents

Chapter 1. From Coin Flipping to Motion of Molecules

Chapter 1

From Coin Flipping to Motion of Molecules

1.1 Introduction

Down among the atoms and molecules randomness rules. Gas molecules rush around in all directions with different speeds, their tracks jagged, zigzag, full of turns and reverses. Disordered. Random. Unpredictable.

Sometimes molecules combine to form ragged structures: branching trees, scraggly nerve cells, patchy forests, wandering coastlines. Jagged. Ragged. Unpredictable.

Other times molecules combine to form structures that are ordered and even: crystals, snowflakes, seashells, human bodies. Orderly. Symmetric. Predictable.

What is science good for? For description! To tell us what is happening. To describe the randomness at the bottom of nature. For prediction! To tell us what is going to happen. To predict the order that grows from randomness.

How does order grow? In large part that is what this whole book is about. While working through it, you will experience chaos, randomness, unpredictability. Nobody in the world can say with certainty what will happen next. Famous people? Powerful people? It doesn't matter: None of them can tell us exactly what will happen next.

Yet you will watch order grow out of disorder. Small numbers of atoms or molecules act unpredictably, move randomly, zigzag. But when large numbers combine, they sometimes result in ordered, predictable patterns. In later chapters you will learn new ways to measure and describe these patterns. But this chapter begins with randomness, the random flipping of coins as a model for the random motion of molecules. And we will begin to see the order that grows from the randomness of both coin-flipping and molecular motion.

Questions To Think About

(1) Jeff is a camp counselor in charge of 20 children. He wants to divide the children into two teams for a game. Each team should have 10 children. Jeff has decided to flip a coin for each camper: heads, the camper goes to one team; tails, go to the other team. Will the children end up evenly divided, ten on each team? Could all the children end up on one team? Which of these results is more likely? Can you think of any way for Jeff to use coin flipping to distribute the children equally into two teams?

(2) You thrust one end of a cold iron poker into a fire. Three minutes later a thermometer half-way along the poker registers a 10-degree rise in temperature. How long do you think it will take for the "cold" end of the poker to rise 10 degrees in temperature?

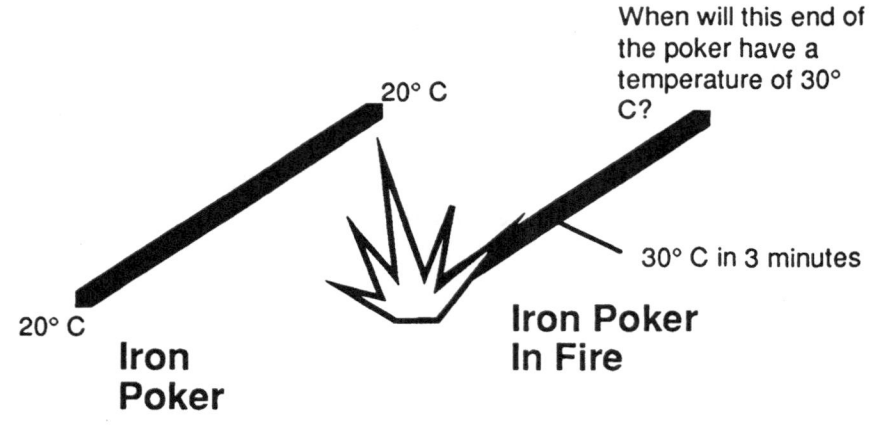

By the time you finish this chapter, you will know the surprising answers to the questions in the box, and you will learn that the two situations are closely related to each other. In the first case flipping a coin is a random way of deciding which team you will be on. In the second case the "hotter" molecules part way along the poker share their increased motion randomly with molecules to the right and to the left—as if the direction of spread of this increased motion were determined by flipping a coin. But how can coin flipping be used to predict the spread of heat, and how are the two cases related to each other? To solve this puzzle, you'll need to know some important things about random motion.

1.2. Randomness: Making Predictions Even When No One Knows What Will Happen Next

What does "random" mean?

Break into small groups of 4 or 5 and discuss the following statements students have made about the word "random." Which of these statements are true and which are false, and why? Choose one person in the group to record and summarize your discussion. After five minutes reassemble and let each group report back to the class.

Flipping a coin *is not* random, because there are only two ways it can come out.

Rolling a 6-sided die *is not* random, because there are only six ways it can come out.

Whether or not I win the state Lottery *is* random, because there are so many people playing the Lottery at the same time.

The weather *is* random, because there are so many conditions that affect the weather.

Activity 1.1 Lottery Game

Is the present always influenced by the past? Suppose you are flipping a coin and, by chance, flip three heads in a row. Does three heads in a row affect the next flip—the fourth flip—or not? Is the fourth flip more likely to be another head? Or is the fourth flip less likely to be a head?

1. Do you believe in "winning streaks": Three heads in a
   row means the next flip is *more* likely to be a head? *Why*
   do you believe in winning streaks?
2. Or do you expect your "luck to run out": Three heads in a
   row means the next flip is *less* likely to be a head? *Why*
   do you believe that your luck can run out?
3. Or do you expect the fourth flip to give a head or a tail
   with equal likelihood, independent of what went before?
   Why do you believe that the next flip is independent?

Try it out with real coins! Flip a coin over and over again until
you get three heads in a row. Now choose one of these strategies
and stick to it.

Strategy #1: If you believe in winning streaks, bet that the
next flip will be a head.
Strategy #2: If you believe in luck running out, bet that the
next flip will be a tail.
Strategy #3: If you believe the next flip is random, bet on
heads the first time, bet on tails the next, and so forth.

Now start flipping the coin again until you get three heads in a
row, then make another bet, using the same strategy. Assume
that a win brings you $1.00 and a loss costs you $1.00. Keep track
of how much "money" you win or lose.

> Can we speed this up? Suppose you flip three different coins at once by
> shaking them in your cupped hands and throwing them on the table. Then
> just look to see if all three are heads. If not, shake them up and throw
> them down again and again until all three do come up heads. Then flip a
> fourth coin by hand and see if you win or lose according to your strategy
> above. Does this give the same result as flipping a fourth time after the
> same coin has come up three times in a row?

Now move on to the computer program called **Lottery**. This
time the computer program allows you to bet on the outcome of
a coin-flip after *four* coins in a row land on the same side (four
heads or four tails). With the computer flipping coins, you can
test your strategy faster than you could flipping coins by hand.
What is the result?

Is your strategy a winner? or a loser? Or do you break even? If
others in the class are doing the same activity, pool your results
in order to compare the success rates of different strategies.

How do we know that the computer is not cheating us? that it is really flipping the coins honestly? Can you think of a test that you could carry out to determine the answers to these questions?

1.3. How Often Something Happens (Probability Distributions)

No one can say with certainty what happens next in a random process such as a coin flip. Even so, some kinds of predictions are possible and useful. If you flip 10 coins, about how many of them do you expect to come up heads? Is it possible that all 10 will come up heads? Is 10 heads in a row *likely*? Is it possible that all 10 will come up tails? Is 10 tails in a row *likely*? One way to answer these questions is to try it out. Let's flip some coins!

What Do You Think?

One thousand students are each given one penny. Each student flips his or her penny ten times and records the number of heads. How many out of 1000 students will flip 10 heads, 9 heads, 8 heads, etc.? *Guess* the answers to these questions and fill in the following table. Be prepared to describe the reasoning behind your guesses.

No. of Heads	No. of Students (out of 1000)
10	_ _ _ _
9	_ _ _ _
8	_ _ _ _
7	_ _ _ _
6	_ _ _ _
5	_ _ _ _
4	_ _ _ _
3	_ _ _ _
2	_ _ _ _
1	_ _ _ _
0	_ _ _ _

Sketch your prediction by drawing a bar graph or histogram on a copy of the following figure. Add your own scale to the vertical axis.

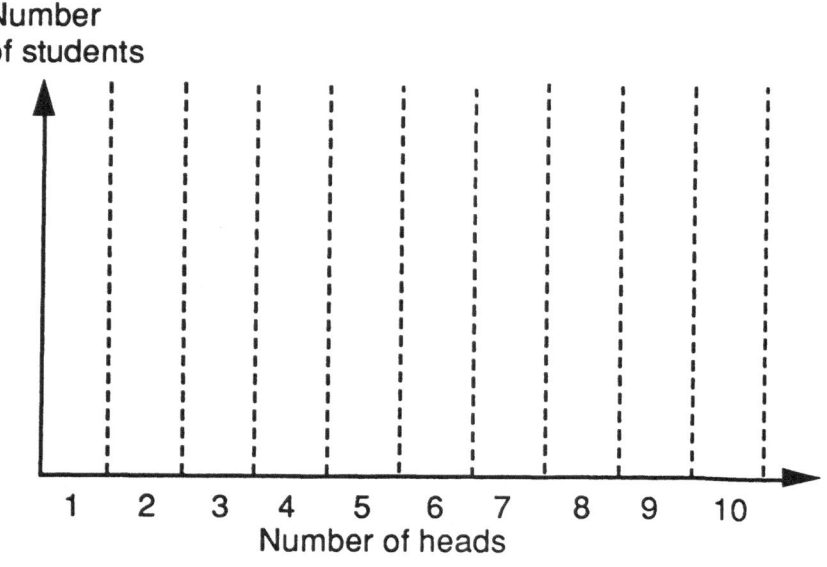

Activity 1.2 Coin Flipping

Now let's find out what is the real outcome of flipping a coin ten times. Work in teams of two or three. Save time by "flipping" ten coins at once: Shake up ten coins between your cupped hands and throw them on the table. Count the number of heads and report this number to the team member who keeps a tally. Shake up the ten coins and drop them again. Again count the number of heads and report this number. Do this ten times.

Now combine the results for all the teams in the class. Have someone make a graph like the following on the board. Put a big X for each trial, stacking each X on top of the previous one for that number of heads, as in the histogram of Figure 1-1.

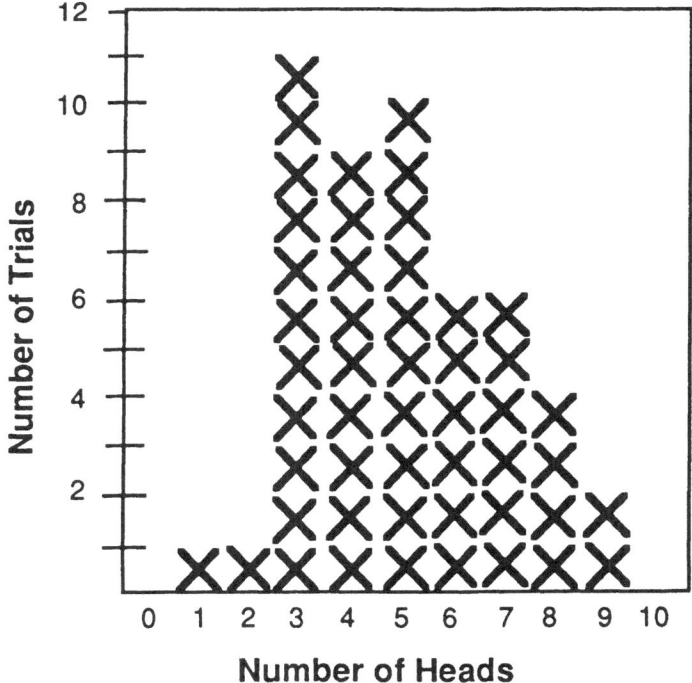

Figure 1-1 One possible histogram of the number of heads when 10 coins are flipped in 50 trials.

Meet with your team and discuss the results plotted on the board. Does the tally look like what you predicted? Why is it uneven in shape? What would the tally look like after 1000 trials? Report back to the class as a whole and discuss your conclusions and questions, including the following:

• Did certain numbers of heads occur more often than others?

• What number of heads is the most likely? What fraction of the time does that actually happen?

• Were there any trials that resulted in zero heads or ten heads?

• Why is the distribution lopsided? If we did this again, could the new distribution be lopsided the other way? Why or why not?

• Based on this activity, would you predict the same histogram for 1000 trials as you guessed earlier? If not, what would you predict now for 1000 trials?

Activity 1.3 Flipping Coins by Computer

Computers may not be intelligent, but they are quick! A computer can flip coins very fast. **Random Walk** is a program that does our ten-pennies experiment thousands of times faster than we can. The computer also plots the results.

1. Start the Random Walk program, which begins with the coin-flipping game.
2. Click on the **Flip** button to flip one coin at a time.
3. Click **Flip** again. And again. Repeat until you have flipped ten coins.
4. Study the numbers at the bottom and the top of the graph: What is being recorded?
5. Now repeat the process, but this time click **Go** to have the computer flip coins automatically, one after the other. Notice that as the bar graph grows, the vertical scale changes to keep the plot on the screen.
6. Record your results in your notebook.
7. Now choose a new window (choose **New** under the **File** menu).
8. Click **Go** and watch the graph grow.
9. Choose the **Tile Windows** command under the **Options** menu and compare the two graphs. **Tile Windows**

shrinks the windows so you can view more than one at the same time side by side.

Do you observe any patterns? Can you predict the next graph? Try it!

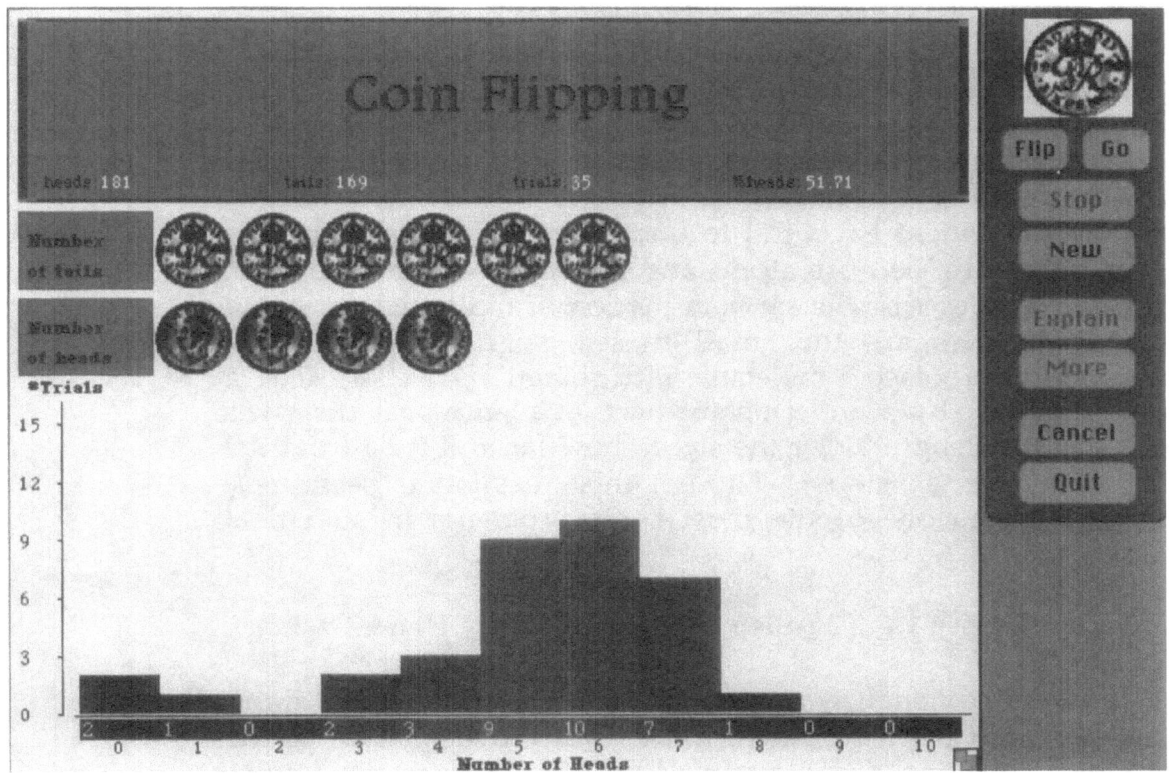

Figure 1-2. Distribution of number of heads in repeated trials flipping 10 coins.

10. Speed up the computer by choosing the **Fast** speed under the **Options** menu. This eliminates the coin display. The first display performs 100 trials of the ten-flip experiment.
11. Select 500 trials under the **Options** menu and then click on **New** in the **File** menu.
13. Select the **Tile Windows** command under **Options** to see all of these results side by side.

As the number of trials increases how does the shape of the graph change? Is the bar graph smoother after 500 trials than after 100 trials?

14. Experiment on your own. Change the number of trials. You can also change the number of flips per trial; this is

called **Number of Steps** under the **Options** menu.
(*Because of a quirk of the program, you must set the Number of Steps and the Number of Trials BEFORE choosing New*.) Before each experiment, make a prediction about what the graph will look like. Does the number of trials or the number of flips affect the shape of the graph? How?

1.4 Random Movement (Random Walks)

In the preceding section we found that some predictability grows out of random coin flipping. This section carries the idea further, relating random coin flipping to random motion. Random movement is central to understanding the microscopic world of Nature, because atoms and molecules move randomly. How can we describe the random motion of molecules in, say, a gas? Molecules are too small to see, so to help us think concretely we replace a molecule with something we can see: a wandering ant. If a wandering ant starts at a lamp post and takes steps of equal length along the street, how far will it be from the lamp post after N steps? Though this question is seemingly trivial, it poses one of the most basic problems in statistical science.

Activity 1.4 Ten-Step Random Walk by Hand

It is easiest to visualize random motion (random walk) along one line, that is, in one dimension.   Call x the position of the ant (i.e. walker) on a one-dimensional line. Locate the origin, that is x = 0, at the lamp post. Then let each "step" of the ant—right or left along the line—be of equal length. One way to picture this is to use one row of a checkerboard or an enlarged version of Figure 1-3.

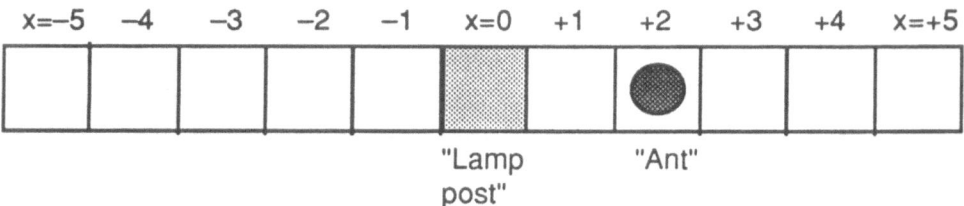

Figure 1-3 Diagram of a wandering ant. How many steps do you think he's taken?

Choose the direction of the step by flipping a penny: If it is a head, the walker steps right and x increases by one; after a tail it steps left and x decreases by one. A head or tail is equally likely; therefore the ant has equal probability of stepping right or left.

Work again with your partner. Use a silver-colored coin (nickel, dime, or quarter) to represent the position of the ant. To begin, put the "ant" in a center cell as the position of the lamp post where the ant starts. The ant steps from one cell to the next, right or left randomly, depending on whether the penny comes up heads or tails, respectively.

1. Flip a penny ten times and move your "ant" accordingly.
2. After ten steps, report the final position of the ant, and whether it is to the right or to the left of the lamp post.
3. Again, the tally keeper puts a big X on a bar graph of position on the blackboard, as you did in Activity 1.2 (Figure 1-1).

How does your result compare with the results of Activity 1.2, in which 10 pennies are flipped at once? Is there a relationship between these two activities? Notice that in this present activity:

$$\begin{pmatrix} \text{final number} \\ \text{of steps to the} \\ \text{right of center} \end{pmatrix} = (\text{number of heads}) - (\text{number of tails})$$

Activity 1.5 Random Walk Program: 1-D Random Walk

Now do the same activity using the computer.

1. Call up the **Random Walk** program and choose **1D Random Walk** under the **Experiment** menu.
2. Take one step at a time by pressing the **Flip** button.
3. Now make the computer flip pennies automatically using the **Go** button.
4. Do the 10-step random walk 10 times.
5. Select **Tile Windows** under the **Options** menu, then select **Fast** under **Options**, then **100 trials** under **Options** and start a new Random Walk of 10 steps.
6. When the 100 trials are finished, do a third 10-step walk 500 times, using the **Fast** setting under **Speed** in the **Options** menu.

Are there any similarities among the bar graphs in the three
runs? Does a larger number of trials make it easier to describe
the shape of the resulting bar graph? If you were shown only the
bar graph, could you tell whether it came from the Random
Walk part of the program or the Coin Flipping part of the
program?

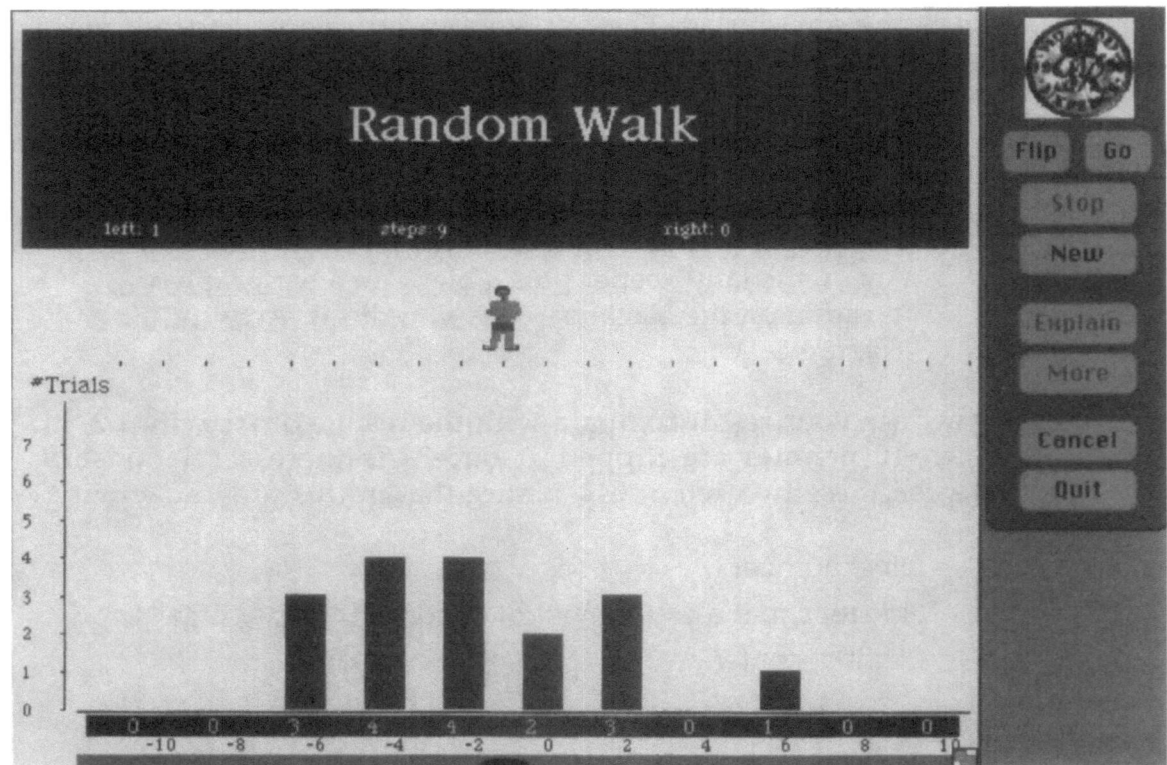

Figure 1-4 Distribution resulting from 10-step random
walk.

1.5 Pascal's Triangle

Pascal's triangle can be pictured as a triangular array of pegs, as shown in Figure 1.5.

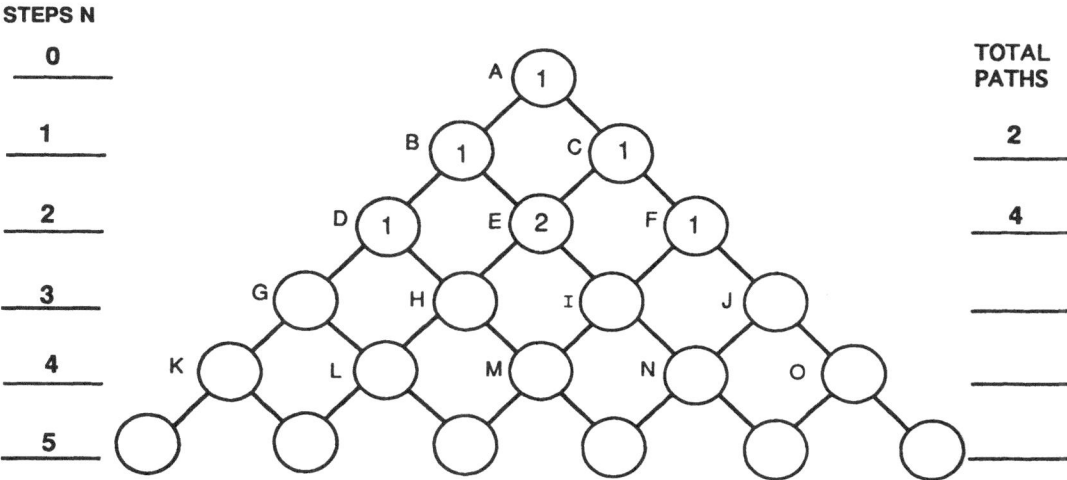

Figure 1-5 Pascal's Triangle of pegs, through which a marble falls

Activity 1.6 Random Walk Program: Pascal's Triangle

Pascal's Triangle is a different means of representing coin flipping or random walks. We are still thinking of a random walk in one dimension (i.e., on a line). Start at the origin — the lamp post — and at every time step flip a coin. If the coin lands as a head, move to the right, else (if it lands as a tail) move to the left. Now represent this process using the triangular array of pegs given by Pascal's triangle, with a ball or marble falling down through the pegs from top to bottom. Flipping a coin and stepping to the right or left is replaced by falling to the right or left of the peg as the ball drops down one level on the triangle. Ten steps means a drop of ten levels.

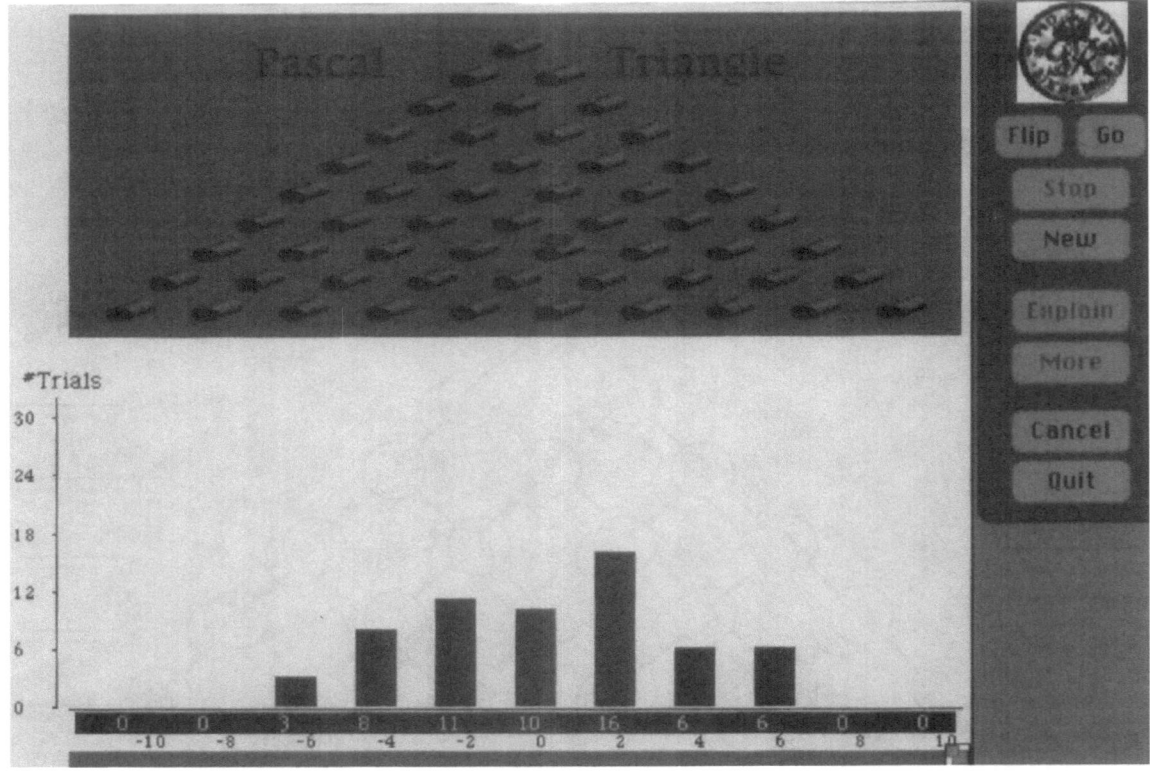

Figure 1-6 Distribution of marbles after falling through Pascal's triangle.

Carry out the following steps with the **Random Walk** program.

1. Choose **Pascal's Triangle** from the Experiment menu
2. Start it doing 100 trials with ten steps. (If you wish, select **Fast** under **Options**.)
3. Select **Tile Windows** under **Options** to place the resulting bar graph in one corner of the screen.
4. Start the coin-flipping experiment, run 100 trials, then display the results in a second tiled window to place it next to the Pascal's Triangle results.
5. Finally call up and start the 1D Random Walk program with 100 trials and show the resulting display in a third small window.

Compare the graphs in the three displays. Are they identical? similar? If you ran any of the programs twice in a row, would you get the same result each time? similar results each time?

How likely is it that a walker will be at least four spaces away from its starting point after taking only four steps? After taking 8 steps? After taking 12 steps? We need to figure out how this likelihood changes as the number of steps changes.

Why is the random walker important? Because it mimics the way a molecules moves. We want to study not only the motion of a single molecule, but also the combined motion of many molecules that define the chemical and physical processes we observe every day. To do this, we need to watch many walkers at the same time.

Activity 1.7 Many Walkers Program: Distribution Width

Instead of watching one walker at a time, let's watch many walkers at the same time. Call up the program **ManyWalkers**. In this program you choose the number of walkers displayed, from one walker to 250 walkers. When you press the STEP button, each of these walkers takes a step randomly, either to the right or to the left. The number of walkers in each position (left or right of the origin) is shown in the bar graph at the bottom of the screen.

Try the display for different numbers of walkers. How are the results different for 40 walkers than for 250 walkers? than for 1 walker? Do more walkers result in a *wider* spread, for the same number of steps? Do more walkers result in a *smoother* graph? Guess: How many walkers do you think it would take to yield a *perfectly smooth* graph?

How does the width of the spread change as the number of steps changes? Is the width after 12 steps three times the width after only 4 steps?

Activity 1.8 Average Position After N Steps

How far from the origin does the wandering ant end up after some number (N) of steps? What do you guess: after 10 steps is the walker more likely to be to the right or to the left of its starting point? If another ant takes 10 steps from the starting point, then another ant, then another ant, what do you expect their <u>average</u> final position to be after 10 steps?

Investigate! Go back to the **ManyWalkers** program and observe the value of "AVG. x" given at the right of the bar graph. The symbol x stands for the displacement of each walkers away from the initial position in units of steps—positive to the right, negative to the left. How does this average change as the number of steps increases? Is AVG. x bigger for more walkers? Or is it smaller for more walkers? Press the SAVE button to record any interesting set of averages; then press the TABLE button to examine the data you have saved.

> **What does theory have to say about the value of the average position of many random walkers?** Here we are not talking about averaging a small number of walkers: not 100 walkers, not 1000 walkers, not even one million walkers. We ask, What is the average position of an *ideally infinite* number of walkers. Answer: The value of the average position is zero, the position of the starting point! How can this be? One word gives the reason: Symmetry! After any fixed number of steps, the walker is equally likely to be to the right (positive displacement) as to the left of the starting point (negative displacement). Moreover, for a given number of steps, the distance from the starting points is likely to have the same value, whether displacement is to the right or to the left. In brief, averaged over many millions of trials, the positive displacements cancel out the negative displacements. Therefore we expect the *average* of many trials to be zero; the average position of the walkers is at the starting point.

The *average* position may be zero, but the *spread* of positions is not zero, as shown also by the example in Figure 1-1. The number of heads in 10 tries does not always come out the same; they are different in different 10-step trials.

How can this spread of final positions be described? Pascal's Triangle helps to answer this question.

Look at the Pascal Triangle shown in Figure 1-5. We are going to do something that looks silly but turns out to be very useful: Count the number of paths that can lead to each of the pegs in the 2nd row, the 3rd row, the 4th row, etc.

The first two rows are done already. For example, the only way to get to point D (after 2 steps) is by going left-left (LL). We enter a 1 in circle D, meaning there is only one way to get there. In contrast, there are two alternative paths to point E, namely left-right (LR) or right-left (RL). Therefore we enter the number 2 in circle E. How many paths are there to point F? Now you fill in the next two rows. You need 3 steps and 4 steps to reach **the pegs** in these rows, respectively.

Now we state a Central Rule of Statistics. Starting from some initial point, and for a random choice at each step. . .

every alternative path to a given final point is equally likely

For example: Look at row 3: There are two paths to central position E, but only one path to each of the end positions D and F. The rule above says that a random walker is twice as likely to end up at the central position E than at either of the positions D or F.

> Is this rule reasonable? We have assumed that as the ball comes to each peg, it is equally likely to fall to the left of the peg as to the right. And every possible path is made up of a sequence of these equal choices. So it is reasonable that every path to a given final point is as likely as every other possible path to that point.

It is easy to find a rule that allows us to determine how many (equally likely!) paths lead to a given peg in Pascal's Triangle: The number in each circle is the sum of the numbers in the two adjacent circles above it in the previous row. (For the end circles, the number is the same as in the one adjacent circle in the row above.) For example, the number in circle E is 2, the sum of the numbers 1+ 1 in circles B and C above it. Reasonable? Think of paths. There is 1 possible path leading into B, one possible path leading into C. Therefore they provide:

$$1 + 1 = 2 \text{ possible paths}$$

to circle E below them.

Use this rule to check your entries in the circles of rows 3 and 4. Also put numbers in the circles for the fifth row at the bottom of Figure 1-5.

Using Pascal's Triangle, we have described the number of different paths (and relative likelihood, that is, probability) of arriving at any location (at any circle in the diagram) after a certain number of steps. The result is too many numbers. It is time to simplify our picture by using averages.

1.6 Average Displacement and Average Squared Displacement

Starting at the lamp post, the ant wanders a certain number of steps, randomly to the right and to the left. The ant records its final separation from the lamp post: positive separation measured to the right, negative separation measured to the left. Then the ant goes back to the lamp post and takes the same number of random steps again, for a second trial, recording its final position. Then a third trial, then a fourth trial, and so on. Finally the ant tries to figure out the average final separation from the lamp post; we average over all trials. Question: What do you expect this average separation to be?

> We already know the answer: This average final position is zero, the starting point. This answer is verified by the numbers in the circles of Pascal's Triangle (Figure 1-5). For every row (each row representing the expected displacements after a given number of steps) the numbers in each circle is the same to the right of the initial position (positive final displacement) as to the left of the initial position (negative final displacement). In taking the average, final displacements to the right are typically canceled by final displacements to the left.

Thus zero is the average displacement of the random walker after many trials, no matter how many steps the walker takes. Yet we know that the *spread* of final positions increases with the number of steps. The number of final positions available increases as the number of steps increases in Pascal's triangle. That is why the triangle is wider at the bottom. Notice that after two steps, 2 of the total 4 possibilities (1 + 2 + 1) leave the ant at its starting point. That's 50%. After 4 steps, we have 6 of the total 16 possibilities at the starting point—a decrease to 37.5%. After 8 steps, what percentage of the possibilities leaves the ant at its starting point? What other calculations can we consider which will help us to understand the spread in final positions?

There are various ways to measure this spread. We would like to get around the fact that rightward and leftward displacements tend to cancel one another. One possibility is to treat as positive both leftward final positions and rightward final positions. That is, forget the minus sign for leftward displacements. Another way to say this is that we average the *absolute values* of the displacements. An absolute number is always positive;

therefore when we average the absolute displacements, we will get a result that is always positive. This leads to the idea of an *average absolute displacement*.

> Return to the **ManyWalkers** program. This time pay attention to the value of "AVG. |x|" given at the right of the bar graph. The symbol |x| stands for "absolute value of x," or "magnitude of x." Does this value increase with the number of steps? For a given number of steps, does AVG. |x| have a larger value for more walkers?

The average absolute displacement is not the measure chosen by scientists to describe the random walker, for a reason given below. Actually it turns out to be simpler to take the *square* of each final displacement and then average these squares. The square of a real number is always positive, even when the number itself is negative. Therefore the average of squares of final displacement will always be positive This average is called the *average squared displacement* or *mean square displacement*.

Here is a sample result of one experiment where 20 ants each took 3 steps:

Final displacement	Number of ants
− 3	2
− 1	9
1	7
3	2

To find the average square displacement we calculate as follows:

2 ants had a square displacement of $(-3)^2$ or 9

9 ants had a square displacement of $(-1)^2$ or 1

7 ants had a square displacement of $(1)^2$ or 1

3 ants had a square displacement of $(3)^2$ or 9

Averaging we get $\dfrac{[\,2(9) + 9(1) + 7(1) + 2(9)\,]}{20}$ = 2.6 for the average squared displacement. (The average number of steps does not have to be an integer; if I take one step and you take two steps, the average is 1.5 steps.)

This result will naturally be a bit different for each trial.

Return to your coins.

1. Start with the "walker" in the center.
2. Flip a coin and move your walker one step.
3. Record its position (+1 or −1) in the table below.
4. Now flip the coin again and record the new position.
5. Continue for a total of five steps, recording each position.
6. Now square the displacement after each step.
7. We wish to graph the average squared displacement versus the number of steps. Plot your data in a distinctive color, say red, on a graph with STEPS on the horizontal axis and x^2 on the vertical.

STEP	Walker One		Walker Two			Walker Three		
	$x =$	$x^2=$	$x =$	$x^2=$	Avg x^2 of walkers #1 &#2	$x =$	$x^2=$	Avg x^2 of walkers #1,2 &3
1								
2								
3								
4								
5								

8. Repeat steps 1 through 7 with a second walker, again recording the position after each step.
9. This time take the <u>average</u> of the squared displacements of the <u>two</u> walkers and plot this in another color (green?) on the graph.
10. Continue with the third walker, this time taking the average of the squared displacements of all three walkers after each step. Plot this in yet another color (blue?).

As more walkers are added, does the graph of the resulting averages approach a pattern ? Add a gray line to the graph to show this pattern.

Activity 1.9 Measures of Average Squared Displacement

Can we make any prediction about the value of the **average squared displacement** after many trials? Once more, we can get lots of examples from the computer.

1. Call up the **ManyWalkers** and look at the value of "AVG. x^2" at the right of the bar graph.
2. Try different numbers of walkers and different numbers of steps.

Does AVG. x^2 increase with number of *walkers*, for a given number of *steps*? Or, in contrast, does AVG. x^2 increase with the number of *steps*, for a given number of *walkers*?

3. SAVE representative data to the table and examine the TABLE.
4. Call up the GRAPH.

Do you notice anything which might help you predict the value of AVG. x^2 ? In particular, can you predict the value of AVG. x^2 for 250 walkers after 50 steps?

Appendix A has a more formal analysis of average squared displacement.

1.7 Diffusion: Forward Motion from Random Motion

1.7.1 The Wandering Ant — What Do You Think?

Suppose that the ant can move in any direction away from the lamp post, not just along a line. This is called a 2-dimensional random walk.

For example, an ant is standing in the center of a 11 by 11 grid, as shown in Figure 1-7. Each grid square is the size of one step. The ant can move one step at a time in one of four directions: north, south, east, or west. The ant cannot move diagonally or take more than one step at a time. If the ant walks off the edge of the grid, it cannot return.

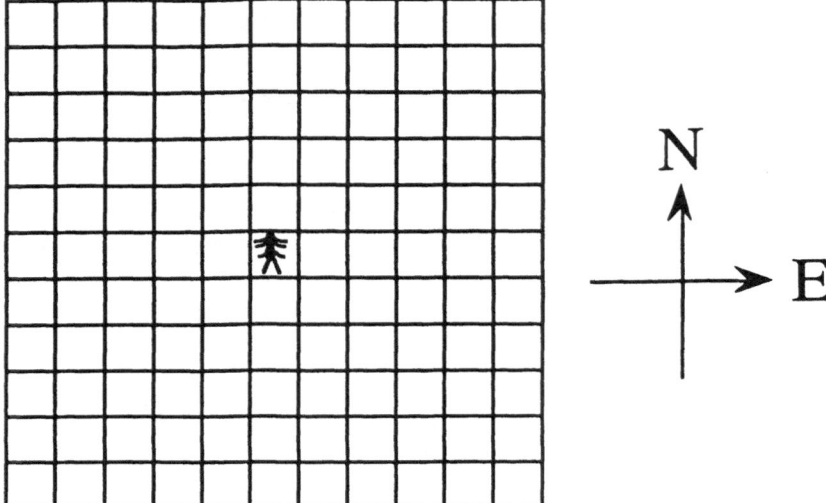

Figure 1-7 The wandering ant.

Discuss the following questions in small groups and report back to the class.

1. Where do you think the ant will most likely be after 10 steps? Will it still be on the grid?

2. Where do you think the ant will most likely be after 100 steps? Will it still be on the grid?

3. Let's say we place 1000 ants on the center square of the grid. If each ant moves independently using the same rules as above, how do you think the ants will be distributed on the grid after 10 steps? After 100 steps? After 1000 steps?

4. Is there a relation between random walks and coin flipping? If so, what?

Activity 1.10 Hands-on Random Walk in 2-Dimensions

Place a checker in the center of a checkerboard. Flip a four sided die labeled north, south, east and west and move the "ant" accordingly. After 10 steps, plot on a lattice the final position of the random walker. Start again from the center and repeat the same procedure ten times. Measure the *distance* from the origin for each random walker and take the average. If several groups are doing the same activity, average your averages. What is your result? Compute the *square* of the distance from the origin for all walkers and take the average? What is your result?

Activity 1.11 Computer Random Walk in 2-Dimensions

Start the program **Anthill**. From the menu, choose one random walker. Plot the average displacement and the average squared displacement as a function of time. What are your results? How do your results change as you increase the number of walkers? How do your results compare to the one-dimensional case?

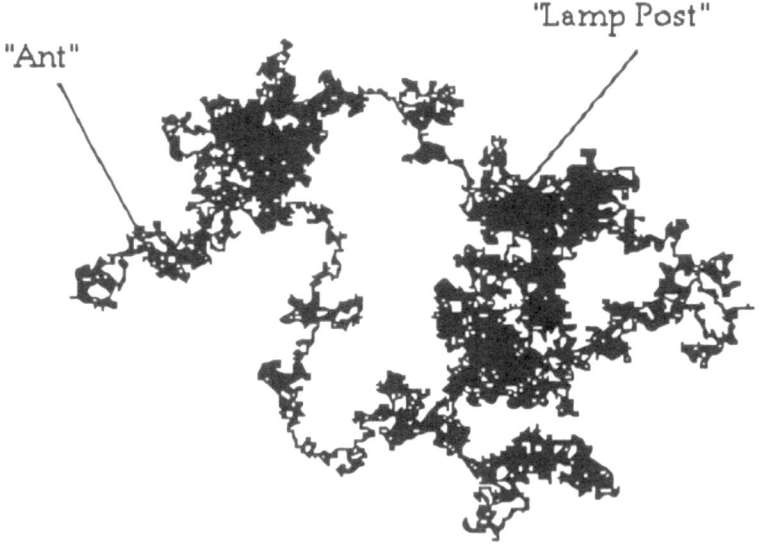

"Ant"   "Lamp Post"

Figure 1-8 Screen shot of the trajectory of 50,000 steps of a single random walker in two dimensions.

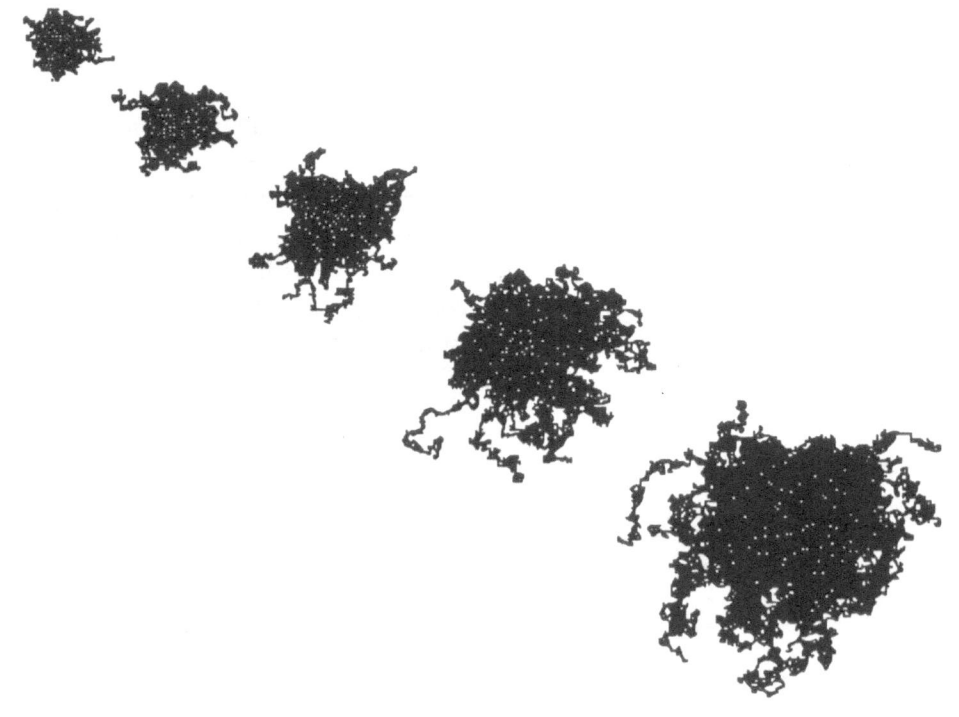

Figure 1-9 Screen shots of the trajectory of 100 random walkers after each has taken 100, 200, 400, 800, and 1600 steps. Can you measure the mean squared displacement from this figure?

Activity 1-12 Number of Distinct Sites Visited

The number of "new" (or distinct) sites visited by a random walker after t steps is of great interest, as it provides a direct measure of the territory covered by a diffusing particle. A new site is defined as any site in space that has not been previously visited by *any* random walker. This quantity appears in the description of many phenomena of interest in ecology, metallurgy, chemistry, and physics.

Two areas of active investigation in random walks involves calculating the time (in steps) for a single walker to hit another walker or to achieve some other goal, such as inducing a reaction by reaching a specific site. Another active area involves calculating the behavior of *many* random walkers. New territory is added as sites are visited for the first time. The problem becomes complicated, because only the first walker to visit a new site counts, so we keep track of the past history of all the walkers.

Consider a colony of termites let loose from the center of a cell filled with sand (Section 2.9). As the termites move, they tunnel through the sand in search of food which is located at the ends of the cell. What can we say about the trails left behind?

Another example: A colony of muskrats is found and their locations recorded by an anthropologist. As the years go by, the anthropologist keeps track of these muskrats and plots contours of the position of the muskrats. What can we say about the contours?

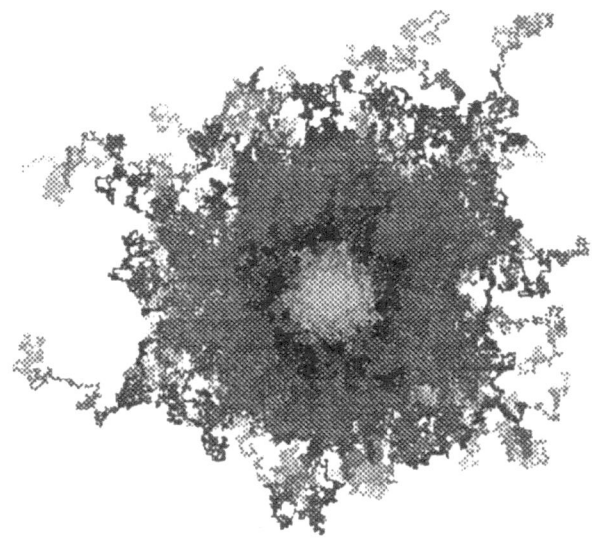

Figure 1-10 Computer model of the contours of multiple particle diffusion. Here, 250 particles are launched from the center. As particles diffuse outward, the color changes so that initially the particles are clumped in the center (indicated by the circular light region in the center). The perimeter becomes rough as particles explore the boundary.

The generalization of the number of distinct sites visited by large numbers of random walkers is particularly relevant to a range of problems—for example, the problem in mathematical ecology of defining the territory covered by members of a given species.

Using Anthill, start many ants in the center of the screen and notice the shape of the "eaten" area as the time progresses. Can you describe the "roughness" of the surface as a function of time?

Activity 1-13. Biased Random Walk: Approximating Nature

If the ant has unequal probability to move in the four directions, we say that the random walk is **biased**. For example, suppose the ant has a probability p=1/2 of going north and probability p=1/4 to go west, p=1/4 to go east, and *zero* probability p=0 of going south. In this case, the walkers never go south and the territory covered moves north quite quickly. Often, nature can be quite biased. Consider the way the smell of an uncorked bottle of ammonia travels from one end of the room to the other (Section 1.7.3). The smell is carried by air currents. In this example, one would have to wait a very long time to smell the ammonia if the molecules were truly diffusing. In a draft, the motion of these "ammonia particles" are quite biased.

Can you list examples from nature where the random motion is biased? And can you list some of the possible sources of the bias (e.g., draft air currents would bias the smell of the pie).

Bias the walkers by allowing them a higher probability to go in a particular direction. Observe the shape of the "eaten" area.

1.7.2 Population Dynamics
Another example of a 2-dimensional random walk is deer moving about a field as they graze. Deer eat, mature, and have baby deer. All the deer depend on grass for food. But what if too many deer are born, too much grass is eaten before it can grow back? Will some deer starve? Will the population of deer reach a constant size? get smaller? die out? The **Deer** program is designed to help you answer these questions.

The **Deer** program is a simple model of population dynamics. Each deer (red square) is a 2-dimensional random walker that wanders around a large field. When the deer enters a green square, it eats the grass in that square. The grass in that square grows back only after a certain time. In the meanwhile, another deer stepping into that square has nothing to eat. A deer will die after a certain number of steps without food.

However, the deer reproduce at a certain age, and each offspring becomes an independent random walker who must eat to survive, just like all the other deer. In the **Deer** program you control the number of steps a deer can live without food, the time it takes for grass in a square to grow back after being eaten, and the breeding age of the deer.

Activity 1.14 Population Dynamics: The Deer Program

Deer Program: Trial Run #1

1. Call up the **Deer** program.
2. After reading the introduction, select **Start** under the **Control** menu. In the upper left you see a green field.
3. Create a single deer by clicking the cursor once in the middle of the field.
4. Start the program running by clicking outside the field.
5. The graphs at the right report the number of deer and the percentage of grass that has been eaten.
6. When the graphs get near to the right side, stop the program.

Answer the following questions:

What happens to the size of your deer population? Does it grow and grow? Does it increase, then decrease? Does it die out?

Predict: Suppose you started again with a deer in exactly the same position. Would the population at every step of the second run be exactly the same as the first? Would the final result be the same? Try it!

Deer Program: Trial Run #2

Now you are going to place a dozen or so deer on the field initially and then let the program run as before. Before doing this predict what will happen:

Will the initial population growth be different for 12 initial deer than for one deer? Will the final outcome be different for 12 initial deer than for one deer?

Now start the program as before, but this time click at a dozen different points on the field before starting the program by clicking outside the field.

What happens to a deer that wanders off, say, the right-hand side of the field?

What is the result of starting with several deer?   Were your predictions correct?

Deer Program: Research Project

Call up the deer program again.   This time select **Change Parameters** from the **Control** menu.   You will see the window shown in Figure 1-11.

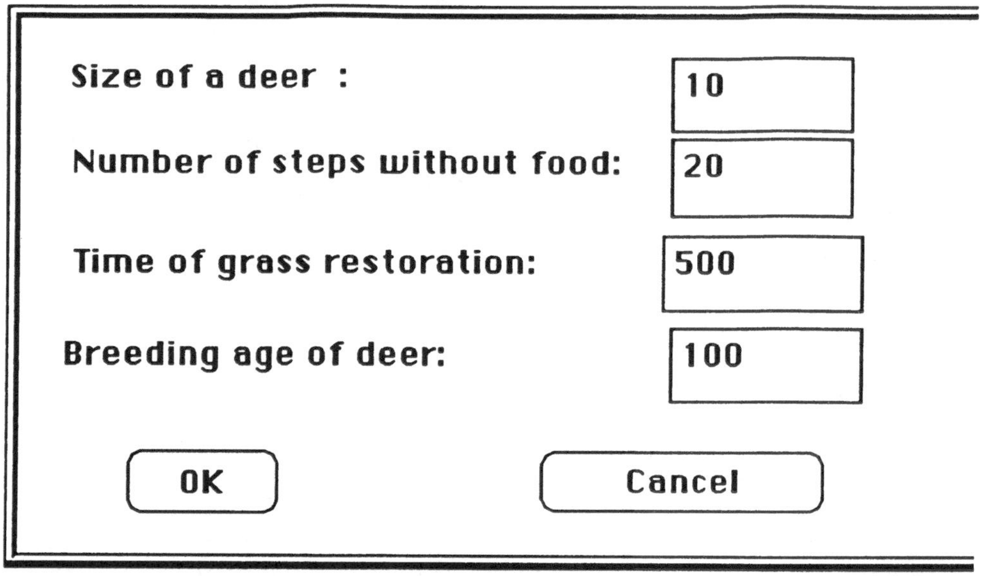

Size of a deer :   10

Number of steps without food: 20

Time of grass restoration: 500

Breeding age of deer: 100

  OK   Cancel

Figure 1-11 The Change Parameters window of the Deer program

The **Size of a deer** determines how big the field is: Smaller deer means smaller squares on a field of constant size, and so a field that can accommodate more steps, more grass plots, and more deer.

A deer dies if it fails to step on a green square for a number of steps greater than the setting of the **Number of steps without food**.

The time and age settings refer to the number of step-cycles that the program has been through.   In one step-cycle every deer takes one random step.

The **Time of grass restoration** is the number of step-cycles required before grass is restored to an "eaten" square.

The **Breeding age of deer** is the number of step-cycles after birth at which a deer has one offspring. Every deer has one (and only one) offspring—provided it does not starve before it reaches this breeding age.

Your first task is to find a combination of these settings such that the population of deer reaches a constant value and stays there—more or less! This is called a **stable population**. Work in groups. Be systematic about the search, writing down in your notebook each setting and the population outcome.

Assemble a "Research Conference" in which different groups report to the class what combinations of settings lead to stable populations. Is there more than one combination of settings that achieve this goal? If so, is there a pattern of such settings?

Your second task is to find which changes in the settings from this stable condition results in an increase in population and which changes result in a decrease in population. Is there a pattern? What changes lead to an oscillating population, one that increases and decreases rhythmically or erratically?

Assemble a second "Research Conference" in which different groups report their results and the class as a whole discusses the conditions for different kinds of population changes. The conference should also discuss the following question:

How would adding a predator complicate the problem and change the outcome? A predator eats deer, has baby predators, and dies if it does not meet a deer after a given number of steps.

1.7.3 Jennifer Runs Out of the Classroom
(**To the teacher:** Start this as a class discussion, encouraging speculation by students and writing these speculations on the board. At the beginning encourage inventiveness rather than accuracy. Only after this is done should the random walk nature of gas be introduced.)

In one corner of the laboratory, Barry opens a bottle of ammonia. Jennifer, who is sitting in the opposite corner of the classroom, grabs her nose and shouts, "Barry, close that bottle!" Barry does so, but Jennifer continues to smell the ammonia, so she runs out of the classroom.

How does the smell get from the opened bottle to Jennifer, 10 meters away? Perhaps air currents in the room carry the smell to her nose. But if there were no air currents—if the air were perfectly still—could the smell still make it to Jennifer's nose?

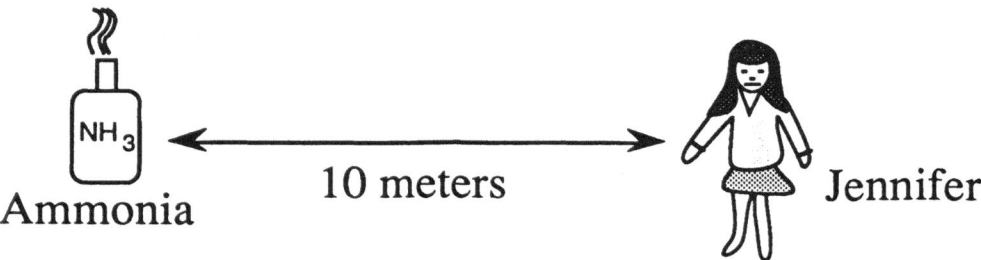

Figure 1-12 When does Jennifer smell the ammonia?

Class discussion.
Write down on the board answers to the following questions given by different members of the class. Add your own questions and possible answers to your questions too.

Start with a quick guess: How soon will Jennifer smell the ammonia in still air?

What is carrying the odor to Jennifer? What moves through the air?

Whatever moves through the air: What path does it take? Straight line? jagged path? circles?

Very rough estimate
Suppose the ammonia molecules do a random walk. That is, each molecule moves in a straight line until it collides with an air molecule, and this collision changes its direction of motion randomly without changing its average energy of motion. This process is called **diffusion**. Of course, diffusion of air molecules is 3-dimensional, but begin by analyzing it assuming one dimensional motion.

From the **ManyWalkers** program we know that in a one-dimensional random walk the *square* of the average distance traveled is proportional to the number of steps. Let x^2 stand for this square average, let N_{steps} be the number of steps, and let L_{step} be the average length of each step, the average distance traveled between collisions. Then the results of **ManyWalkers** tells us that:

$$x = (\text{average distance travelled}) = \sqrt{\overline{x^2}} = \sqrt{N_{steps}} \; L_{step}$$

or

$$N_{step} = \frac{x^2}{L_{step}^2}$$

Notice that for uniform motion in a straight line, the formula would be $N = x/L$ (without the squares).

How many steps will the ammonia molecule take to get across the room? To begin to answer this question we need to know the speed of the molecules at room temperature and the average distance that a molecule moves between collisions. Here are approximate values for these two quantities. You will derive these values later.

1. **Speed of gas molecules.** A gas molecule at room temperature moves with an average speed approximately equal to that of a rifle bullet, about 600 meters/second. Why doesn't this hurt? This speed is approximately 2200 kilometers/hour or 1300 miles/hour. Why don't you feel a 1300 mph wind all the time?

2. **Distance between collisions.** A gas molecule in the air travels a distance between collisions (the length of one step) approximately equal to $L_{step} = 1000 \times 3 \times 10^{-10}$ meters $= 3 \times 10^{-7}$ meters. How does this distance compare with the diameter of one hairs on your head? This distance is equal to how many times the diameter of an atom?

From these quantities and earlier equations, answer the following questions:

1. Suppose that the molecule travels in a straight line. How many steps N_{step} would it take for the molecule to travel $x = 10$ meters? Compare this number with the number of people who live in your country.

2. Suppose that the molecule travels in a one-dimensional random walk. How many steps (collisions) N_{step} will it

take to travel x = 10 meters? Compare this number with
the number of people alive on Earth.

3. How long a time t_{step}, on the average, elapses between
    collisions?

4. Therefore how many seconds t are required for the
    molecule to move 10 meters, on the average? Express
    your answer in an everyday unit, such as hours or days or
    years.

5. What if it is one meter instead of 10 meters. Don't work
    too hard. Find an equation that can give you the answer
    in one step.

Fill in the steps of your result by estimating two numbers given
above, the speed of gas molecules and the distance between
collisions.

1. Speed of gas molecules.
The average kinetic energy of a molecule in a gas at absolute
temperature T is approximately (3/2) kT, where k is Boltzmann's
constant (1.381 x 10^{-23} joules/^{o}K).

$$\frac{3}{2}kT = \frac{1}{2}mv^2$$

From this, find an expression for the average speed of a gas
molecule.

2. Distance between collisions
A gas molecule will experience a collision if the distance from its
center to another molecule is equal to the sum of their two radii,
or approximately equal to the *diameter* d of a single molecule.
We can think of a moving molecule as sweeping out a volume
in one second which is equal to its velocity v (meters/second)
times the area A = π d^2, where d is this diameter. See Figure 1-
13.

$$\frac{Volume}{sec\,ond} = v\pi d^2$$

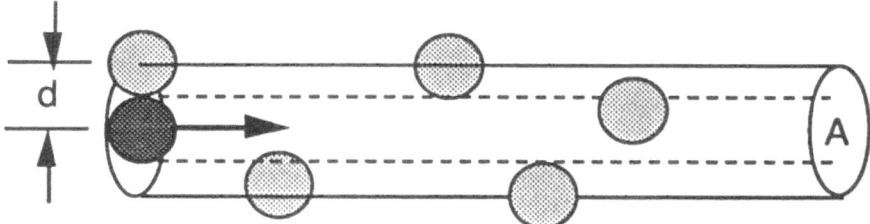

Figure 1-13 A moving molecule collides with any other molecule whose center lies within a cylinder of radius equal to the twice the radius of the molecules.

The moving molecule will collide with any other molecule whose center lies in this volume. (Of course whenever the molecule hits another molecule, it changes direction. So the cylinder shown in the figure will be broken up into many short segments. But the volume swept out will still be given approximately by the expression above.)

How many gas molecules are to be found with their centers inside this volume V, on the average? Let the density of gas molecules be n (molecules/meter^3). Then the number of molecules to be found in this volume, equal to the number of collisions per second, is given by

$$(\text{collisions / second}) = \frac{nV}{\text{second}} = nv\pi d^2$$

Substitute values to find the number of collisions per second.

From this result and the speed of the molecule, find an approximate value for the distance L_{step} the molecule travels between collisions.

The values for v and L_{step} you derived here may be different from the values you used to estimate the time t for the ammonia smell to cross the x = 10-meter room. Recalculate the time t using your new values.

Activity 1.15 Diffusion Chamber Experiment

What does it all mean? Someone might say, "Who cares about the fate of a wandering ant or a grazing deer or an ammonia molecule? What does all this have to do with real science?" Answer: A lot! On a microscopic level, a molecule moves randomly, zig-zags as it bumps into other molecules, moves away from some initial point in the same way the wandering ant staggers away from the lamp post or the deer grazes around the field. This random-walk from one position to another position is called **diffusion**. In a laboratory, we cannot possibly keep track of how each molecule moves, because the molecule is too small and there are too many molecules. However, we can imitate the motion of each molecule with a random walk and look at average results. Such average motion is much more predictable than the zig-zag motion of a single molecule.

A popular chemistry demonstration is the **diffusion chamber**, in which two different gases, typically NH_3: ammonia (molecular weight 17) and HCl: hydrogen chloride (molecular weight 36) diffuse from opposite ends of a long closed glass tube. Eventually the two gases meet and react, forming a disk of white dust NH_4Cl: ammonium chloride part way along the tube, *not* at the middle.

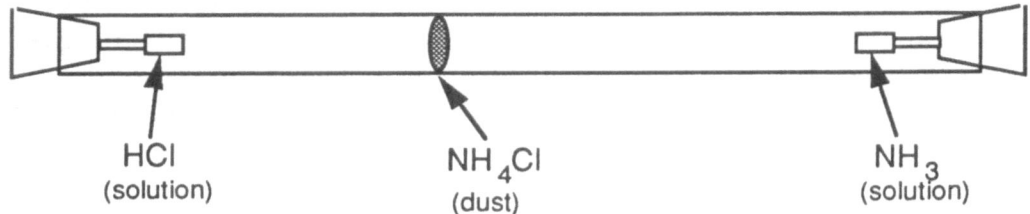

HCl
(solution)

NH$_4$Cl
(dust)

NH$_3$
(solution)

Figure 1-14 Diffusion Chamber Experiment. The acid hydrogen chloride HCl (in water) is placed at one end, and evaporates as HCl gas. Ammonia, which is a water solution of ammonium hydroxide, is placed at the other end, and gives off ammonia gas, NH_3. The two gases diffuse down the tube. Where these gases first meet, they react chemically to make a disk-shaped cloud of dust composed of (the solid) ammonium chloride NH_4Cl.

CAUTION: SOME OF THESE CHEMICALS ARE DANGEROUS. LET YOUR TEACHER DO THE EXPERIMENT AS A DEMONSTRATION.

Before carrying out the experiment, predict where along the tube the disk will form and estimate how long it will take before the disk appears, using the following outlines or some other methods.

Position of the disk

Where along the tube will the dust disk appear? Jennifer and John now get into an argument. They have very different opinions as to where the disk of dust should form, at least based on their respective understandings of molecular kinetics.

John argues that it is the speed of the molecules which determines the position at which the disk of dust forms. He says: "The kinetic energies of the ammonia molecules and the hydrogen chloride molecules are, on average, the same. That's the case when the temperature in a room, or at least in a local volume, is uniform. Since the kinetic energy K of a molecule is given by:

$$K = \frac{1}{2}mv^2$$

we must have

$$\frac{1}{2}m_{HCl}v_{HCl}^2 = \frac{1}{2}m_{NH_3}v_{NH_3}^2$$

From this he concludes that the ratio of velocities is:

$$\frac{v_{NH_3}}{v_{HCl}} = \left(\frac{m_{HCl}}{m_{NH_3}}\right)^{1/2}$$

He now says: the disk will form when the ammonia and hydrogen chloride molecules first meet. John says that the result will be an immediate snow of precipitating salt. The time at which they meet is, say, t. But at time t the distance of the lead ammonia molecules from their starting points is

$$x_{NH_3} = v_{NH_3}t$$

while the position of the lead HCl molecules is

$$x_{HCl} = v_{HCl}t$$

The ratio of these positions is:

$$\frac{x_{NH_3}}{x_{HCl}} = \frac{v_{NH_3}}{v_{HCl}} = \left(\frac{m_{HCl}}{m_{NH_3}}\right)^{1/2}$$

For John this is now an easy problem. The molecular mass of ammonia is 17, and that of HCl is 36. On this basis he predicts the ratio of distances they will travel to be 0.69.

Jennifer doesn't agree. She says that all the collisions the ammonia and hydrogen chloride molecules have with the air molecules will change the position of the disk. John sticks to his guns. The speed of the molecules must determine the relative positions he says, and even if the average speed may be "sort of" slowed down by collisions, the final ratio will remain the same.

Jennifer is having some trouble expressing her argument. She complains that John's view of the motion of the HCl and NH_3 molecules is a picture of BBs or marbles moving straight to the disk. Her mental picture is of a cloud gradually advancing with a sort of resistance from the air. Moreover, she thinks that some molecules get ahead of others. It's like the picture of walkers advancing in the ManyWalkers simulation. This makes for some difficulty however. If the gas molecules advance that way, then the concentration of the HCl and NH_3 won't be uniform; Jennifer thinks that there will be a few molecules out in front, with a gradual increase as the rest catch up.

But how to calculate where the disk will appear if the density of the colliding clouds isn't constant? With some uncertainty Jennifer decides to use the relationship that

$$x^2 = N_{step}L^2_{step}$$

Taking a deep breath, Jennifer now says:

The number of collisions per second is $nv\pi d^2$, so in time t the number of steps is $N_{step} = nv\pi d^2 t$.

Above Jennifer also found that the average length of each step has the value:

$$L_{step} = \frac{v}{nv\pi d^2} = \frac{1}{n\pi d^2}$$

Combining these results she reasons that

$$x^2 = \frac{nv\pi d^2 t}{\left(n\pi d^2\right)^2} = \frac{vt}{n\pi d^2} = vL_{step}t$$

Taking a ratio similar to John's, Jennifer has:

$$\frac{x^2_{NH_3}}{x^2_{HCl}} = \frac{v_{NH_3}L_{NH_3}}{v_{HCl}L_{HCl}}$$

and finally

$$\frac{x_{NH_3}}{x_{HCl}} = \left(\frac{v_{NH_3}L_{NH_3}}{v_{HCl}L_{HCl}}\right)^{1/2}$$

Jennifer is a bit uneasy about this result. First, she took the square root of square averages — is that the right measure of where the two gases will first make contact? And how to measure L_{NH3} and L_{HCl}? She needs d, the "diameter" for a linear molecule (HCl), and for a trigonal molecule (NH_3). What does this mean anyway? She could average over the rotated molecule, but that makes her head spin. Suppose that she approximates L_{HCl} and L_{NH3} as equal. After all, most of the collisions which determine these values are with air molecules — at least these are identical for both types of collisions. If she makes this drastic approximation she finds that:

$$\frac{x_{NH_3}}{x_{HCl}} = \left(\frac{v_{NH_3}}{v_{HCl}}\right)^{1/2} = \left(\frac{m_{HCl}}{m_{NH_3}}\right)^{1/4}$$

Jennifer finds that the ratio of positions goes as the 1/4 power of the mass ratio, while John found that it goes as the 1/2 power. Numerically, Jennifer predicts that:

$$\frac{x_{NH_3}}{x_{HCl}} = \left(\frac{36}{17}\right)^{1/4} = 0.83$$

This result is qualitatively and quantitatively different from John's. As for its quantitative accuracy, Jennifer is completely uncertain since she had to equate the mean step lengths of the two different molecules.

Time for formation of the disk

From the story of Jennifer and the ammonia bottle, estimate how long a time it will take for the disk to appear. Do not do elaborate calculations: Use an earlier result to obtain your estimate from a single equation.

Experiment: If possible, carry out the diffusion experiment in diffusion chambers of two different lengths, starting with the short one. Record the time from the start of the experiment until the disk of white dust becomes visible and the position of the disk. Next predict the time for the disk to become visible in a diffusion chamber that is twice as long as the first. Carry out the experiment. Is the result what you predicted?

Now call up the **Diffusion Chamber** program and model the two experiments twice, once for a short length, once for twice that length. How much longer does it take for 10 dust molecules to form in the long chamber than the short chamber? Twice as long? Ten times as long? Do each model twice: Once for straight-line motion of the molecules (called "ballistic"), once for random motion. Which kind of motion is the better model for the real experiment?

Back to the Poker

Now go back and answer again the "Poker Problem" in the box at the beginning of this chapter. Does it take twice as long (or four times as long or nine times as long or what?) for heat to travel the whole length of the poker as it does to travel half the length?

1.8 Periodic Precipitation: Liesegang Rings

The diffusion chamber experiment is an example of what happens when diffusion is coupled to a chemical reaction. In the case of the ammonia and hydrogen chloride, the reaction resulted in the precipitation of ammonium chloride. More broadly, we refer to this as an example of pattern formation by **reaction-diffusion.**

In the diffusion tube experiment the pattern you viewed was simple: a disk of ammonium chloride. However, reaction-diffusion processes have been linked to the patterns observed in agates, the stripes of the zebra, and the development of embryos.

In the experiment described here you will have the opportunity to study a more complex pattern emerging from a reaction-diffusion system. Just as the combination of ammonia and hydrogen chloride results in a precipitate, so does the combination of potassium chromate and copper sulfate in solution, or in a gel. In this case it is copper chromate which is the insoluble resultant.

In your experiment you will fill a vertical vial with a homogeneous gel containing potassium chromate. On top of the gel you will place copper sulfate crystals. The gel has sufficient water content so that these crystals dissolve on the surface. The dissolved copper sulfate will diffuse into gel.

There are clear similarities and differences between this experiment and the diffusion tube experiment. In both experiments: 1) the reaction is dependent on the diffusion of one of the reactants; 2) the reaction is between two ions resulting in precipitation of a salt; and, 3) in both reactions the geometry is that of diffusion along a cylinder.

There are, however, some apparent differences: 1) in this experiment one of the reactants, the potassium chromate, is initially uniformly distributed along the cylinder; and, 2) the diffusion is occurring in a gel as opposed to air. Can you think of other differences and similarities?

What do you think will happen as the copper sulfate diffuses into the gel? Will a precipitate of copper chromate appear? If one does, what will that do to the local concentration of chromate ion? How will charge neutrality be maintained? Will the concentration of chromate ions remain uniform? Will the concentration of potassium ions remain uniform?

Considering these questions, can you predict the pattern which will emerge in the vial? Does the geometry of the experiment limit the nature of the patterns you might expect? (Scientists would say that the experiment is being conducted with the **symmetry** of a cylinder. Do arguments of **symmetry** help you narrow the range of patterns you might expect?)

Can you make a guess as to how long you may have to wait to observe a pattern? (Hint: the diffusion constant for copper sulfate in the gel is roughly what it is in water, on the order of 10^{-6} cm^2/sec.

Activity 1.16 Liesegang Experiment

1. Following the Appendix, prepare your vial for the experiment.

2. Observe the pattern form. Find the time it takes for portions of the pattern to develop. Measure the distances between portions of the pattern.

Data Analysis

This experiment is largely a qualitative exercise. One simple quantitative question: did the pattern take as long to develop as you expected?

Qualitatively, compare your predicted pattern with what you observed. Is the symmetry what you expected? Can you understand the pattern on the base of the diffusion tube experiment plus the differences between the two set-ups?

Can you use the ManyWalkers program to understand your results?

1.9 Postlude: The Meaning of Models

What Do You Think about Models?

Remarkably, we can make correct predictions about diffusion using our random walk models, even though some assumptions of this model are wrong:

- The random walker in nature does NOT take steps of equal length.
- The random walker in nature does NOT step at equal time intervals.
- The random walker in nature does NOT move only along a single line.

But that's research! That's the way models operate: Models simplify reality so we can start to analyze it. When a model leads to partial success in predicting behavior, then we try to modify and improve the assumptions of the model in order to make more accurate predictions. (Usually this leads to a more complicated model.)

This model business -- what does it really mean? Which of the following do you think is a scientific model? If it is a scientific model, give an example of how it might be used in research.

- Crash-test dummies for automobile testing.
- Laboratory mice.
- Chemical equations
- Computer simulations
- Random Walker computer program

In this chapter we have used a very simple model: an ant wandering back and forth with steps of equal length taken at equal time intervals. Yet this simple model describes many processes in the real world. How can this be, since our model is so simple? Very similar results are predicted by more complicated models that add more randomness: steps of random length, steps in random directions, steps that take place randomly in time. It turns out that the predictions of these more complicated models are similar to ours as long as we use in our model the average step length, average time between steps, and average distance from the starting point. Often in science a simple, easily understood model makes good predictions about the complicated real world.

Chapter 1, Appendix A: Demonstration that Average Square Displacement is Equal to Number of Steps Taken

What is the value of the average square displacement after N steps? The word "average" in "average square" tells us we want the average over many trials. Here we show two ways to calculate this average square displacement.

First method: average square Displacement from Pascal's Triangle

Look back at Pascal's Triangle, Figure 1-2, repeated below as Figure 1-15.

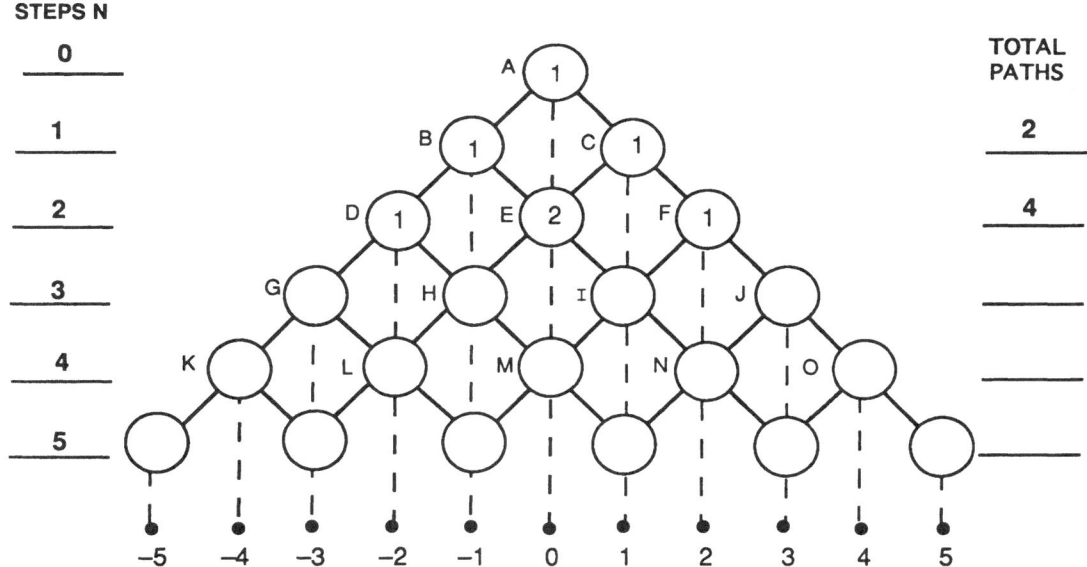

Figure 1-15 Pascal's triangle, repeated.

Pick out the row of circles representing displacements after two steps. A total of 4 paths enter this row (the denominator of the average in the equation below). Two paths lead to zero displacement, one path leads to +2 steps of displacement, and one path leads to –2 steps of displacement. Each of these paths is equally likely. In taking the average of the squares (numerator of the equation below), there is one entry for $(-2)^2 = 4$, two entries for $(0)^2 = 0$, and one entry for $(+2)^2$. Therefore the average of the square for two steps is:

$$\frac{1\,(-2)^2 + 2\,(0)^2 + 1\,(+2)^2}{4} = \frac{8}{4} = 2$$

Now calculate the average square displacement after three steps. A total of 8 paths enter this row. Three paths lead to +1 final displacement. Three paths lead to −1 final displacement. One path leads to +3 and one path to −3 displacement. Follow the same steps as above to calculate the average square displacement after three steps:

$$\frac{1\,(-3)^2 + 3\,(-1)^2 + 3\,(+1)^2 + 1\,(+3)^2}{8} = \frac{24}{8} = 3$$

Show that for the row representing displacement after four steps the average square displacement is 4. What is the average square displacement after five steps?

Do you see a pattern? The average of the square of the final displacement is equal to the number of steps. A graph of the ideal mean square displacement then is simply the line $y = x$. This convenient result is one reason why mathematicians and scientists have chosen to pay attention to the average square displacement.

Second method: Calculating the Average Square Displacement using Algebra
Notation: Scientists often use the symbol x to represent displacement along a line, x^2 to represent the square of this displacement, a subscript N to represent "after N steps," and a bracket < > to represent average value. Then our argument from Pascal's triangle is that:

(average of the squared displacement after N steps) = number of steps N

which is written

$$< (x_N)^2 > = N$$

We derived this result using Pascal's Triangle. Next we derive the same answer using algebra.

How do we find the value of $< (x_N)^2 >$? Sneak up on the answer starting with a smaller number of steps, call it n. After n steps we determine how far we are from the lamp post. We call this value x_n (the displacement) and we square it (i.e., multiply it by itself) to get $(x_n)^2$. [Why do we square it? Because we want to know how far the walker has gone but do not care about the direction; squaring erases the difference between positive and negative directions of motion.] Then we start again and take the same number, n steps, find a second value for x_n and square it to get a second value for $(x_n)^2$. Do this again and again. Finally average all the values we have obtained for $(x_n)^2$. Using carot brackets <> as a symbol for "average," we are finding the value of $\langle(x_n)^2\rangle$.

What do we expect the value of $\langle(x_n)^2\rangle$ to be? Suppose the walker has taken n steps and is now at position x_n. Where will the walker be at the next step x_{n+1}? If we know where the ant is now (i.e., x_n) then after the next step the ant can be a step to the right or a step to the left;

either at $\quad x_{n+1} = x_n + 1$ (step right) (1a) or at $\quad x_{n+1} = x_n - 1$ (step left) (1b)

Which of these will it be? We cannot say for sure. In a random walk both are equally likely. So we take an average: Square both sides of equations (1a) and (1b) and take the average of the two. Again, use the carot parentheses <> to mean average value. Then :

$$\langle(x_{n+1})^2\rangle = \f(\langle(x_n+1)^2 + (x_n-1)^2\rangle, 2) = \f(\langle x_n^2 + 2x_n + 1 + x_n^2 - 2x_n + 1\rangle, 2)$$

or $\quad \langle(x_{n+1})^2\rangle = \dfrac{2\langle(x_n)^2\rangle + 2}{2}$

or $\quad \langle(x_{n+1})^2\rangle = \langle(x_n)^2\rangle + 1$ $\qquad\qquad$ (2)

What does this equation mean? Start with n = 0, the zeroth step (or no step at all). Then $\langle(x_0)^2\rangle = 0$. For the first step, equation (2) tells us that $\langle(x_1)^2\rangle = \langle(x_0)^2\rangle + 1 = 0 + 1 = 1$, which we knew already without this calculation. From this $\langle(x_2)^2\rangle = \langle(x_1)^2\rangle + 1 = 1 + 1 = 2$ and $\langle(x_3)^2\rangle = 3$ and, in general $\langle(x_N)^2\rangle = N$. The result? The average squared displacement after N steps is simply N:

$$\langle(x_N)^2\rangle = N$$

This is the same result we obtained from studying Pascal's Triangle.

What we stated above is an *ideal* average, the predicted average over an infinite number of trials. For a finite number of trials, for example the average of 250 walkers, this is our "best guess" as to the final squared displacement after N random steps along a line. But because steps are random, the best guess rarely matches actual values.

Chapter 1, Appendix B: Preparation for Liesegang Ring Experiment

Materials for this experiment:

Reagents: 0.5 mole acetic acid solution; 0.5 mole potassium chromate solution; commercial water glass (sodium silicate solution) with specific gravity 1.38–1.42 g/ml; copper sulfate crystals; distilled water.

In addition, a 30 ml vial with screw-on cap (size of vial can be varied) and two 50 ml beakers.

Procedure:

In what follows the assumption is that the experiment will be done in a 30 ml vial. For other sizes, scale the necessary ingredients.   The recipe here will leave about 4 ml free at the top of the vial for the copper sulfate.

1.   Make 12.88 ml of solution by adding 10.88 ml of distilled water to 2 ml of sodium silicate solution. This produces a solution with specific gravity 1.06.

2. Add 6.44ml 0.5 M acetic acid and 6.44ml of 0.5 M potassium chromate to the solution.

3. Pour the solution into the vial. Screw on the cap and mix all the reagents by inverting the vial about 10 times. **Don't shake** the vial — this will introduce air bubbles into the gel and potentially distort the pattern.

4. Leave the gel in a safe place where it will not be moved or shaken. The gel will harden in about one and a half hours.

5. After hardening, add enough copper sulfate crystals to cover the entire top surface area of the gel to a height of 0.5 cm.

6. You will start seeing patterns after roughly 24 hours. The pattern will continue forming for about one and a half weeks to two weeks.

REFERENCE: This version of the Liesegang experiment was adapted from J. E. Forman, *Journal of Chemical Education*, Vol. 67, page 720, 1990

Contents

Chapter 2. Fractals in Nature.
Growing and Measuring Random Fractals

Chapter 2

Fractals in Nature
Growing and Measuring
Random Fractals

What Do You Think?

On the following two pages are several pictures of scraggly structures. What do they have in common? Answer the following questions and be prepared to discuss them.

(1)   How are these objects similar to each other in appearance? In what ways do these objects look different?

(2) There are many ways to rank objects. For example, you can rank different students according to weight. You can also rank the same group of students by height or shoe size or age or hair color. Look at the photographs above and figure out different ways to rank these images. Rank the images them according to as many different criteria as you can think of. How inventive can you be?

(3)   The pictures show very different objects, yet their images look rather similar. Why do you think these objects look so much alike? Do you think it is a coincidence? If it is not a coincidence, what are possible ways that their similarity might come from a common cause?

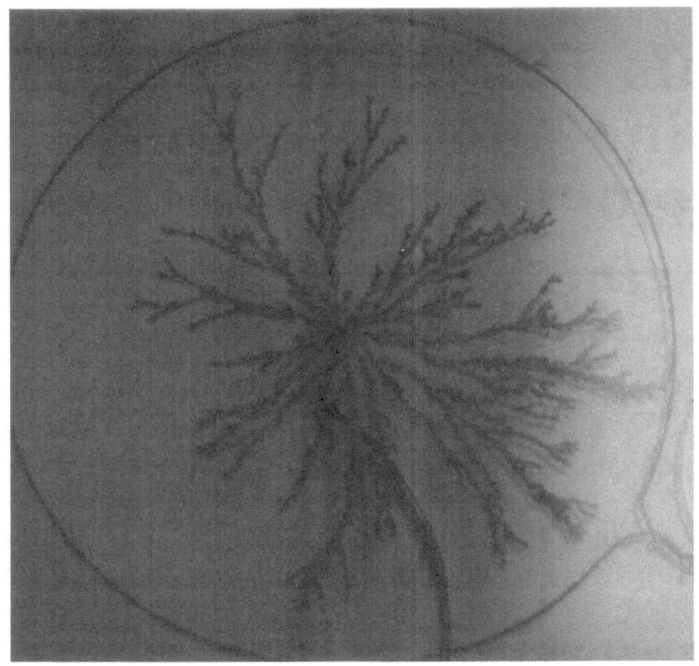

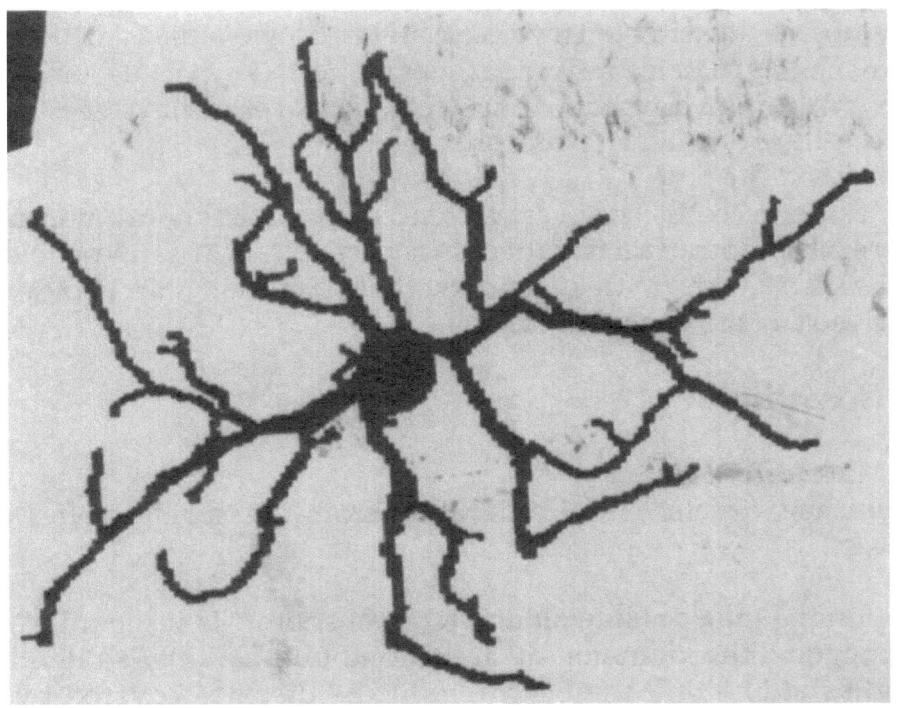

2.1 Introduction

Nature is full of spidery patterns: lightning bolts, coastlines, nerve cells, termite tunnels, bacteria cultures, root systems, forest fires, soil cracking, river deltas, galactic distributions, mountain ranges, tidal patterns, cloud shapes, sequencing of nucleotides in DNA, cauliflower, broccoli, lungs, kidneys, the scraggly nerve cells that carry signals to and from your brain, the branching arteries and veins that make up your circulatory system. These and other similar patterns in nature are called **natural fractals** or **random fractals**. This chapter contains activities that describe random fractals.

> **There are two kinds of fractals:** mathematical fractals and natural (or random) fractals. A **mathematical fractal** can be described by a mathematical formula. Given this formula, the resulting structure is always identically the same (though it may be colored in different ways). In contrast, **natural fractals** never repeat themselves; each one is unique, different from all others. This is because these processes are frequently equivalent to coin-flipping, plus a few simple rules. Nature is full of random fractals. In this book you will explore a few of the many random fractals in Nature.

Branching, scraggly nerve cells are important to life (one of the patterns on the preceding pages). We cannot live without them. How do we describe a nerve cell? How do we classify different nerve cells? Each individual nerve cell is special, unique, different from every other nerve cell. And yet our eye sees that nerve cells are similar to one another.

Besides nerve cells, how can we describe, classify, and measure different random fractal patterns in nature? That is the subject of the activities in this chapter that you will carry out as described in the pages that follow.

2.2 Coastline

2.2.1 Introduction
Measuring a fractal is quite different from measuring circles or squares.

How long is the coastline along an ocean shore? Is it the mileage reading on the odometer of an automobile driving along the straight road behind the beach? Is it the distance covered by a bicyclist who pedals along the shore on a footpath? Is it the length run by a jogger whose morning exercise includes running "next to the water"? Is it the number of steps of a little bird running along the edge of the water looking for washed-up morsels? Whose "length of coastline" is the true one?

Trying to answer these innocent questions leads to deep consequences.

Activity 2.1 "Walking" Along a Coastline

2.2.2 Stepping Along a Coastline with a Caliper
To investigate coastlines, start by "walking" along the map of a coastline provided by your teacher.

1. Obtain from your teacher
   A. a map of the coastline to be measured.
   B. a pair of calipers (dividers). A caliper or divider is like a circle-drawing compass, except that both ends have points.

2. Your teacher will show you
   A. the points on the map between which to measure the length of coastline
   B. the scaled distance on the map for the first setting between the ends of the calipers (preferably a power of two, such as 64 kilometers or 32 miles).

3. Place the first end of the calipers at the beginning of the coastline to be measured and swing your calipers so that the second end rests on the coastline as close to the starting point as possible. This is step number 1. (When you do this, you may "step across" an outward or inward piece of the coastline, Figure 2-1. That's OK. It's what would happen if you really walked a coastline.)

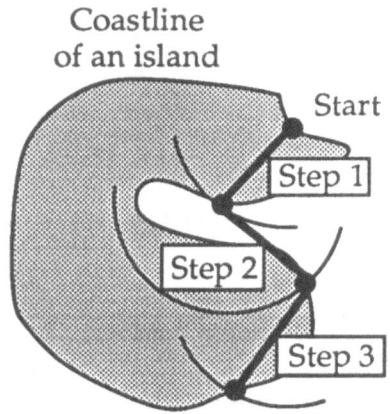

Coastline
of an island

Figure 2-1. Measuring a coastline with caliper.
The caliper often "steps across" a peninsula or inlet.

4. Now swing the caliper so that the first end "walks" along to the nearest point on the coastline. This is step number 2.

5. Continue this process, counting the steps as you go; stop when your next step would carry you beyond the end of the coastline. The number of steps you have taken to this point times the distance covered with each step is one measure of the length of the coastline.

6. Enter the size of the step (scaled size in miles, kilometers, or whatever) in the left column of Table 1. Enter the number of steps in the third column.

7. Now you are going to start again with half the distance between the ends of the calipers that you used the first time. Before you do this, **predict** the number of steps it will take you to measure along the coastline with this smaller step size. Enter your prediction in the second column of the table.

8. "Walk" along the coastline with the calipers set to half the original scale distance. This is equivalent to a child (instead of an adult) walking a shoreline. Count the total number of steps and record the results in the correct columns of Table 2.1. How accurate was your prediction?

9. Repeat the process of cutting the scale distance in half as many times as you can, each time making a prediction, taking the walk, and filling in the table with your results.

10. Fill in the right-hand column of the table by multiplying each step size by the number of steps to get a total "length."

> Did your prediction of number of steps improve as you went along? What happens to the total length of the coastline for different step sizes? **Take a guess:** What is the "true" length of the coastline? How would you define this total length?

Table 2.1. Measuring the length of a coastline

1. Size of step, S	2. PREDICTED Number of Steps, N	3. MEASURED Number of Steps, N	4. "Length" = N x S

Put the map away; you will need it later.

Activity 2.2 Measuring Dimension of a Coastline

2.2.3 A New Kind of Graph Paper

Now you are going to plot on a graph the data you collected from your coastline measurements. A blank piece of graph paper for copying is on the following page. If you have not used such graph paper before, the squished scales may appear odd.

> These scales are explained in Appendix A. For now, you need to know only one thing about the strange scale: You can add a zero to all the numbers along either axis and still have a valid graph paper. For example, the vertical axis can be relabled 10, 20, 40, . . . 100, 200, 400, 1000, 2000, 4000, . . 10,000. The result is perfectly usable. Or you can multiply all numbers along either axis by 1/10. For example the vertical axis can be relabled 0.1, 0.2, 0.4, . . . 1, 2, 4, . . . 10, 20, 40, . . . 100. Still valid graph paper. If the entries in your table do not match the numbers on one axis of the graph paper, just multiply all the numbers on that axis by 10 or by 1/10 until they do match.

1. Plot the numbers from your table on a photocopy of the graph paper. Plot the number of steps along the vertical axis versus the step size along the horizontal axis. What kind of curve do the plotted points seem to follow? Do they lie along a straight line? When they lie along a straight line on this kind of graph, it is

possible to define a **dimension** for the object, equal to the magnitude of the slope of this straight line. For some objects this slope—this dimension—is not a whole number. Objects with non-integer dimension are called *fractals*.

2. Draw by eye the "best" straight line using the points on your graph. Determine the magnitude of the slope of your graph. The slope is defined as

$$Slope = \frac{\text{change in vertical rise along the line}}{\text{change in horizontal run along the line}} = \frac{\text{"rise"}}{\text{"run"}}$$

Use a regular ruler to measure this slope, not the scale along the axes. What is the magnitude of the slope (the dimension) of your coastline?

Save this plot; you will need it in the next activity.

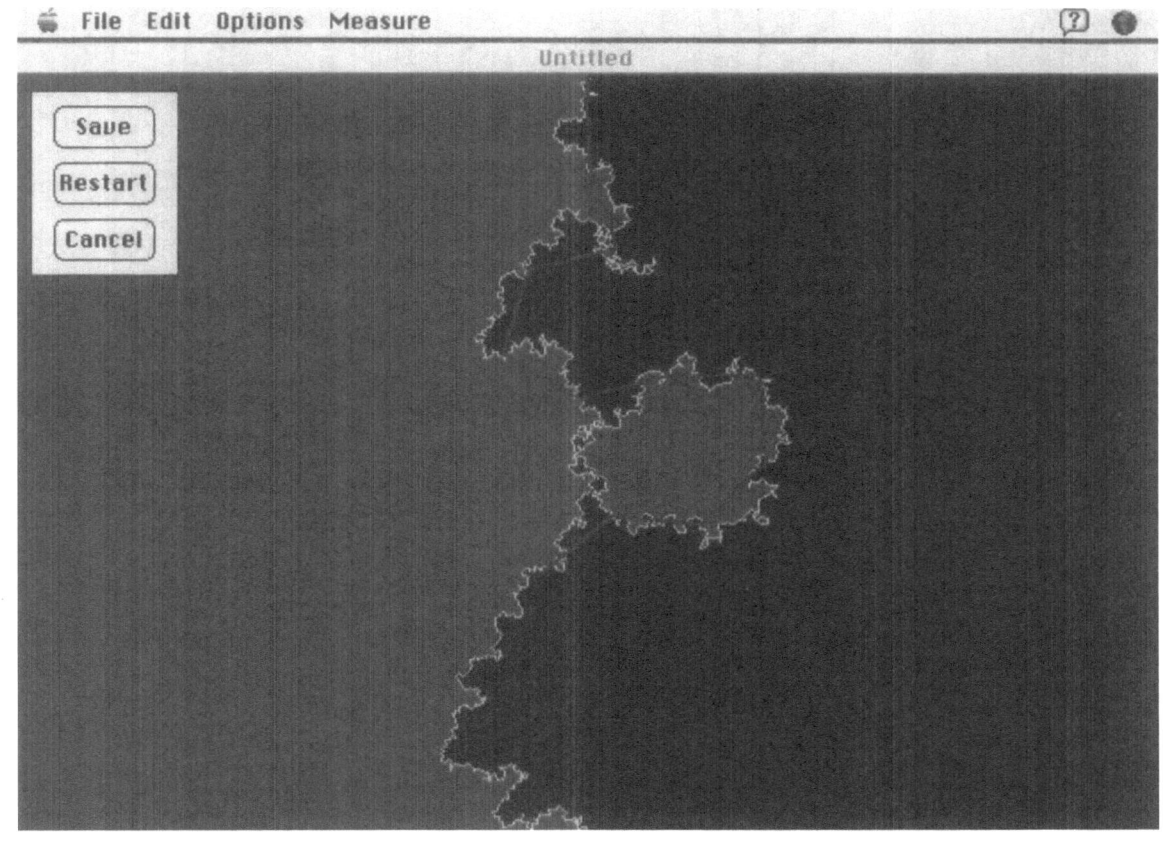

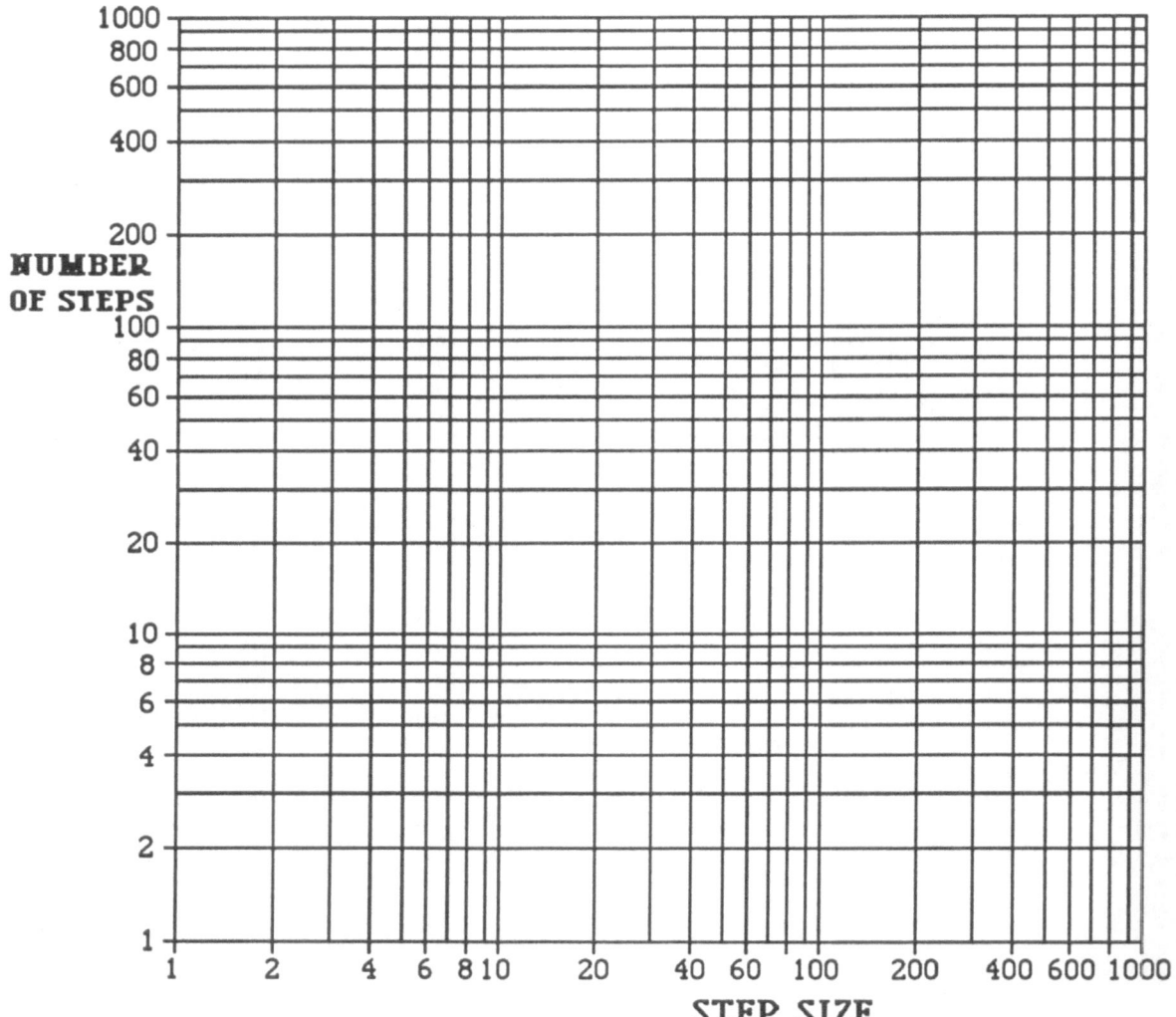

Figure 2-2. Graph paper for plotting the entries in Table 1. Do not draw on this grid. Photocopy it; you will need several copies.

Activity 2.3 Creating Your Own Coastline By Hand

In order to study coastlines further, create a sample coastline, using the following instructions. You will need a rubber band, a ruler, a die, and nine push-pins or thumb tacks.

1. On a bulletin board, stretch the rubber band between two push-pins a vertical distance of 8 inches apart.

2. Locate the midpoint of the rubber band. You are going to pin this midpoint right or left a random amount and hold it there with a third push pin. Here is how:

4. Roll the die.
    *   If you get 1, push the midpoint **left** by **two** inches, hold it with the push pin.
    *   If you get 2, push the midpoint **left** by **one** inch, hold it with the push pin.
    *   If you get 3, leave the midpoint where it is, but hold it with a push pin.
    *   If you get 4, push the midpoint **right** by **one** inch, hold it with the push pin.
    *   If you get 5, push the midpoint **right** by **two** inches, it with the push pin.
    *   If you get 6, roll again.

5. Now look at the two segments which make up your coast. Repeat the above process on each of the segments, pushing the midpoint to one side or the other perpendicular to the direction of the segment. Pin each relocated midpoint to obtain a four-segment coastline.

6. Repeat one more time on the four segments to get an eight-segment coastline. Trace the final coastline on a piece of paper and tape it to the front board.

Activity 2.4 Computer Program for "Walking" Along a Coastline

2.2.4 The Ruler Method for Measuring Dimension

Now we will use a computer program to draw the model of a coastline and to measure the dimension of that model.

Creating a Coastline

1. Open the program **Fractal Coastline**.

2. After the initial screen, click the **New** button. This brings up a control window and a coastline window.

3. Go to the **Options** menu and set the **Animation** to be **Step-by-step,** so that you can watch the construction of the coastline.

4. Click on **OK** to close the **Options** window.

5. The Control window now has a button called **Start Iterations**. Click on this button. This brings up a straight line in the Coastline window.

6. Click on the **Iteration** button once. What rule do you think the computer uses to start the coastline?

7. Click on the **Iteration** button a second time. What does the computer do? Fill in the blanks in the following paragraph.

8. Go on clicking the **Iteration** button (you can do this up to 10 times).

9. When you reach this limit or are satisfied, click the **Done** button.

10. If you want to make additional coastlines more rapidly, go back to the **Animation** command in the **Options** menu and change it back to **Computer Animation.**

11. Make several different coastlines, each time changing the setting on the scroll bar that controls the roughness. Does every roughness setting give you a realistic coastline?

12. Find a roughness setting that gives you a coastline about as jagged as the actual coastline you measured before (even though the overall shape may not be the same). Give this coastline a name using the **Save as . . .** command in the **File** menu.

Measuring the Dimension of a Coastline Using the Ruler Method

1. Now measure the dimension of your constructed coastline. Go to the **Measure** menu. There are two options, one using ruler (calipers) and one using the grid. Choose **Using Ruler**, set a long ruler length with the scroll bar, then select **Manual Measure** from the dialog box.

2. One end of the ruler is anchored at one end of the coastline and the other end points toward the mouse. Move the mouse to set the free end of the ruler at a point on the coastline and click the mouse to anchor it and create a new ruler.

3. Repeat the process until you have stepped along the whole coastline. (If you mess up, press the **Restart** button.)

4. When you are happy with the result, press the "Save" button. (The "Cancel" button takes you back to the ruler control window without saving your measurements.)

5. Change the ruler length, and this time click on the **Automatic Measure** button from the dialog box. Then the computer will do the "stepping" for you.

6. Repeat the process with several different ruler lengths. Each result is automatically saved when you use Automatic Measure.

7. Now let the computer analyze the data. Explore for yourself the "Data . . ." entry in the **Options** menu. This shows the data you have collected and different ways to graph it, similar to what you did by hand in previous activities.

8. The kind of graph we drew in Section 2.2 (Figure 2-2) to measure dimension is called a **LogLog graph.** (See Appendix A for an explanation of the LogLog graph.) Choose the LogLog plot and see the line drawn through the points. In the small window at the right, click on **Curve Fit**; a straight line is drawn through the data points. This line has the equation $y = mx + b$, where m is the slope and the variables y and x are the logarithms of the number of rulers laid down and ruler length respectively. The slope m is the dimension of the coastline, which you measured and calculated manually. Write down the value of this dimension; you will need it later. Does this dimension have a

value approximately equal to the dimension of the actual coastline you measured at the beginning?

9. Be sure you saved this coastline (Step 12 above) under a name you can remember.

Activity 2.5 Covering a Coastline with Boxes by Hand

2.2.5 The Box Method for Measuring Dimension.

Thus far we have used the **ruler method** for measuring the dimension of a coastline. An equivalent method is called the **grid method** or **box method** or **covering method**. The grid method is a bit more general, and can be used for different kinds of fractals. Here is how it works:

1. Get out the map whose coastline you measured originally.

2. Take a second piece of blank paper approximately the same size and cut out from it a square that just covers the entire coastline to be measured. Call the edge-length of this square 16.

3. Now fold the covering square into fourths and cut along the fold-lines.

4. Predict: How many of these 8-wide squares will it take to *cover* the coastline? The word *cover* means that the squares string along the coastline without overlapping, but so that no piece of this coastline is visible. Enter your prediction in the table below.

5. Now carry out the covering with 8-wide squares and enter the result in the table. How good was your prediction?

6. Fold each of the smaller covering squares into equal fourths and cut along the fold lines. Predict how many of these 4-wide squares it will take to cover the coastline. Enter your prediction in the table.

7. Carry out the covering with 4-wide squares and enter the result in the table.

8. Repeat this process with squares of edge-length 2, and 1. In each case predict the number needed to cover the coastline, and

enter the result in the table before measuring and entering the actual covering.

Table 2-2 Covering the Coastline with Boxes

Edge length of square	PREDICT: Number of squares to cover coastline	MEASURE: Number of squares to cover coastline
16	-----------	1
8		
4		
2		
1		

9. Were your predictions closer this time than when you predicted results of the ruler method?

10. Get out your original LogLog graph of the coastline results. Plot the data from this table on the same graph. This time the vertical scale is "number of squares to cover coastline" and the horizontal scale is "edge length of square." Is the result a straight line? If so, is the slope of the line the same as that of the ruler method of measuring this coastline?

11. Based on your use of the ruler and grid methods, which do you find more efficient? more accurate? Explain.

Activity 2.6 Computer Covering of a Coastline

2.2.6 Grid Method with Fractal Coastline Program

Now let the computer help us carry out the grid method quickly and easily.

1. Call up the **Fractal Coastline** Program again.

2. From the **File** menu, select **Open** and call up the Coastline you saved earlier.

3. Open **Animation** menu and click on **Step by Step**.

4. Under the **Measurement** menu, choose **Grid Method**. A window allows you to adjust the size of the grid. (Click on **Preview Grid** to see covering grid.)

5. Click on every grid square in which any part (even a teeny piece) of the coastline shows. When you are done, the resulting set of grid squares should completely cover the coastline. Click on **Save**.

6. Try another grid size. This time under **Animations** change setting to **Automatic Measure** and let the computer cover the coastline. Each result will be saved automatically.

7. Let the computer cover the coastline with grids of a variety of sizes.

8. Now have the computer analyze the data using a LogLog plot to find the dimension of your coastline according to the grid method.

9. Compare the value of the dimension measured using the grid method with the value of the dimension you obtained earlier using the ruler method. (If you did not save the earlier value of the dimension, you can quickly take a few readings using the ruler method now and graph the result.)

10. Save your coastline as a MacPaint file (**File Menu**) under a new name. You will need this file later.

11. Summarize all you have done in this part of the project. Enter values of dimension for both coastlines in the following table:

Table 2-3 Measured Dimensions of Coastline

Measurement method	Dimension of Actual coastline	Dimension of Computer coastline
Ruler/Caliper		
Grid/Box		

2.3 The Meaning of Dimension

2.3.1. One- Two- and Three-Dimensional Objects

We have been measuring the non-integer dimension of coastlines. The shape of a coastline is called fractal. Non-fractal objects have integer dimensions. For example, a line is 1-dimensional. A piece of paper is 2-dimensional. A cube is 3-dimensional, as shown in Figure 2-3.

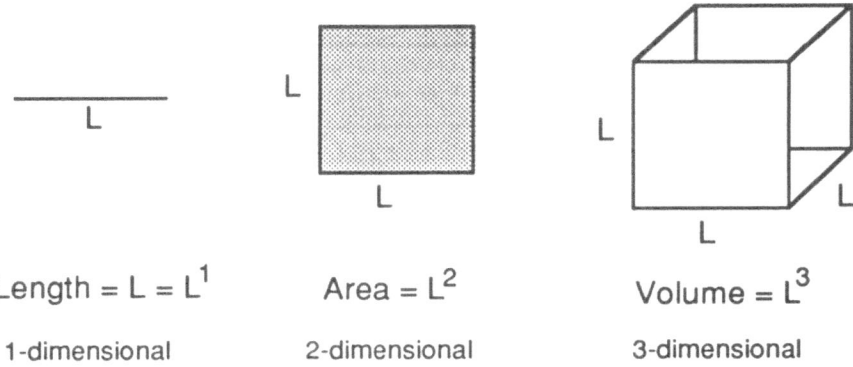

Figure 2-3 The meaning of 1, 2, and 3 dimensions.

Note that the dimension is the exponent in the appropriate measure of the figure—length or area or volume.

2.3.2. Fractal Dimension Using Circles

So far we have been calculating fractal dimension using the method of covering. There are other methods for calculating dimension. One that is useful for objects with dimension less than or equal to two involves measuring the total area of the object inside circles of different radii. For a plane with all squares filled, the area goes up as the square of the radius: $A = \pi R^2$. The power 2 of R tells us—again!—that a disk is 2-dimensional.

Activity 2.7 Hands-on Circle Measurement of Dimension

Check the results of plugging into the formula $A = \pi R^2$ by counting the number of cells inside circles of different radii in Figure 2-4. For each circle, count all the cells inside that circle, not just those between that circle and the next smaller circle. Enter your results in the table on the following page.

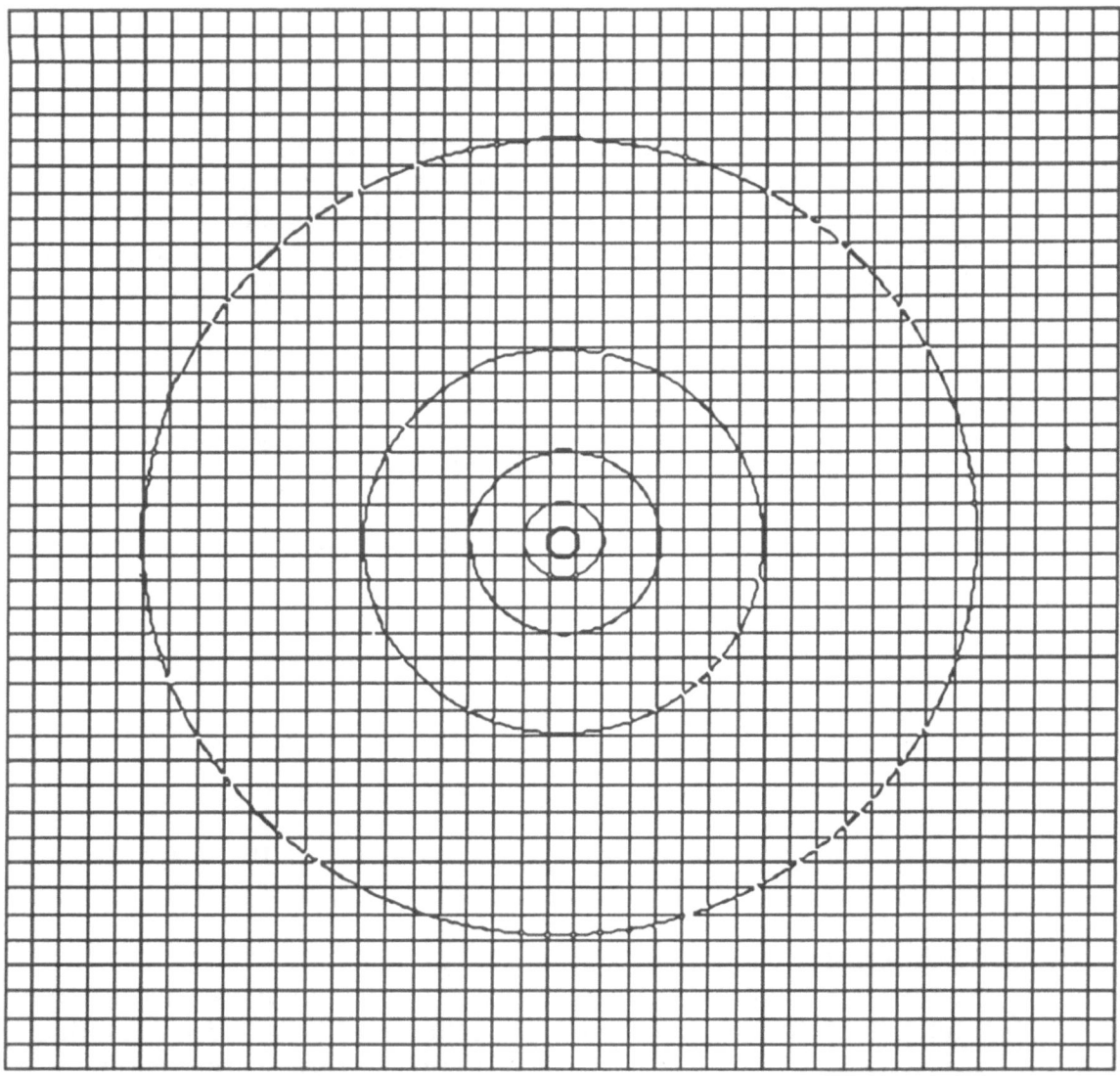

Figure 2-4 Count the cells inside each circle to discover
that the flat surface of the figure is 2-dimensional.
Surprise!

Note: What about *pieces* of cells? If the circle passes through a
cell, should it be counted as inside or outside the circle? You can
decide for yourself, as long as you are consistent. You might
choose any of the following rules, or make up your own:

- If any piece of a cell is inside the circle, the cell is counted as
  if all of it is inside, or

- If any piece of a cell is outside the circle, the cell is counted as if all of it is outside, or

- Count the cells that are part inside, part outside and use half that number in your total count.

Compare your results with someone who used a different rule than you did. Does which rule you use make a difference?

Table 2-4 for use with Figure 2-4

r (radius)	N(Count)
1	
2	
4	
8	
16	

Plot the results in the table on a LogLog graph, the radius scale along the horizontal axis and the count N along the vertical axis. From the measurement of the slope, calculate the dimension of the solid circles.

Now measure the dimension of the line in Figure 2-5 using a similar procedure: First, count the number N of short line segments inside each circle of radius r. Enter the results in Table 2-5. Second, plot N versus r on a LogLog graph and measure the slope. What is the value of the dimension of the line?

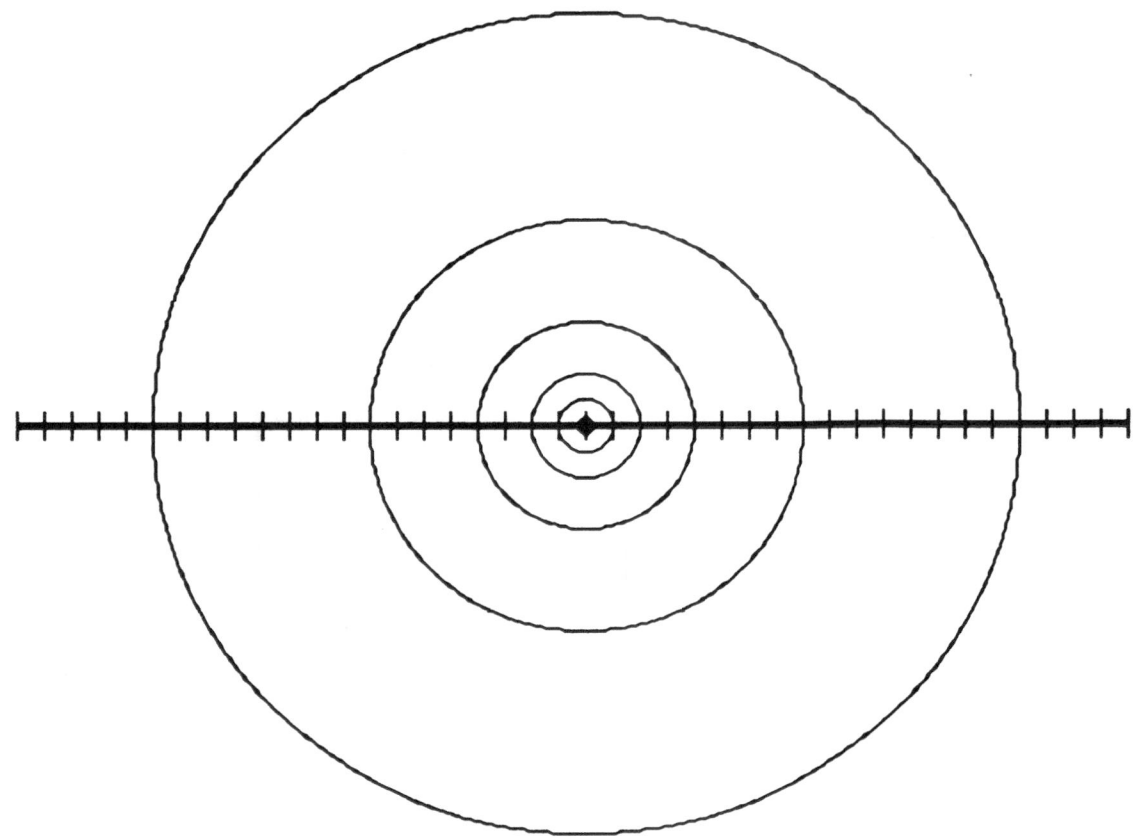

Figure 2-5 Count the number of line segments inside each
circle to discover again that a line is 1-dimensional.

Table 2-5 for use with Figure 2-5

r (radius)	N(Count)
1	
2	
4	
8	
16	

Finally, use the circle method to measure the dimension of the
fractal shown in Figure 2-6 with circles added. This time count
the number N of black squares inside each circle, each with a
different radius r.

Enter your results on a copy of the following table:

Table 2-6 for use with Figure 2-6

r (radius)	N(Count)
1	
2	
4	
8	
16	

The fractal dimension equals the slope of the resulting straight line when N is plotted versus R on a LogLog graph. Carry out this plot and determine the dimension of this fractal?

Question: What is the dimension of a figure like the following? Is this a fractal?

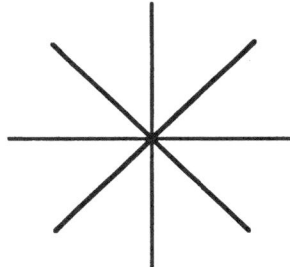

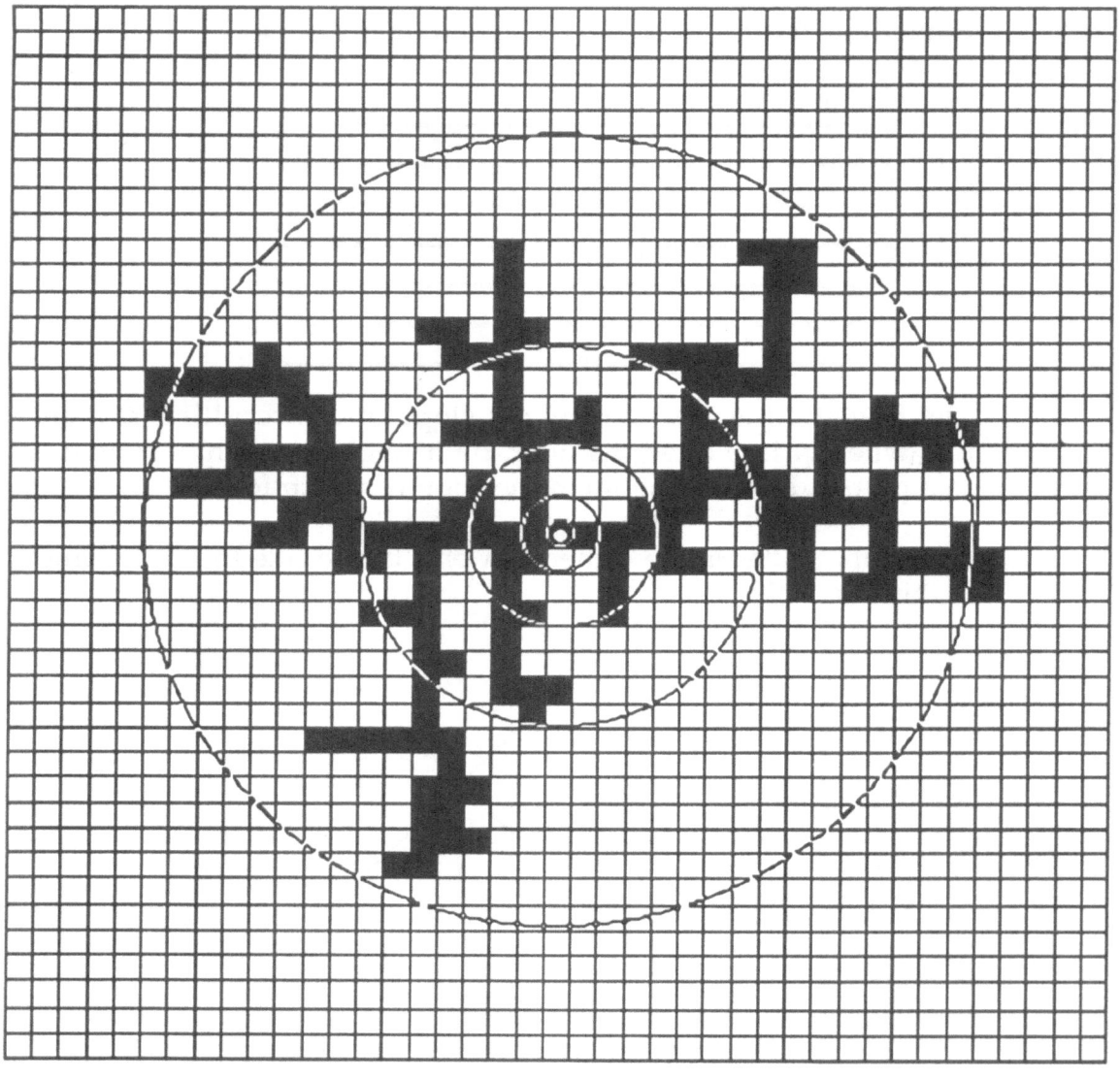

Figure 2-6 Count the number of black cells inside each
circle to discover the dimension of this fractal pattern.

2.4 Growing Random Patterns

2.4.1 Introduction

Is a coastline a *line*? Not really, because a line is one-
dimensional, whereas a coastline is a random fractal with a
dimension between the values 1 and 2. Other random fractal

patterns have dimensions between 2 and 3: a snowflake, a nerve cell, a lightning stroke. These structures can be modeled by a process called **aggregation**. The resulting jagged pattern is called an **aggregate**.

The following set of activities will help you (1) to describe aggregates and random fractals, (2) to model the aggregation process by which the structures grow, and (3) to recognize and measure the basic properties of random fractals.

Activity 2.8 Growing a Pattern in the Laboratory

The shiny chromium surfaces on an expensive automobile are made by dipping the parts into a chemical bath that contains chromium ions and running an electric current through the bath so that chromium metal plates out onto the parts to be coated. This process is called **electroplating** or **electrodeposition**.

The usual goal in electroplating chromium is to produce a smooth surface so the car will look good. In the pattern-forming experiment described below, the result will be quite different from this, as you will see.

2.4.2 Electrochemical Deposition: Classroom Experiment

In the experiment described below you have the opportunity to grow a physical object, and describe its geometry quantitatively by two independent methods. The **electrochemical deposition** experiment is abbreviated **ECD**. The ECD cell (Figure 2-7) consists of a circular anode surrounding a central cathode. The space between the plates is the thickness of the anode wire, about 1/2 mm or 500 microns. Between the plates, and between the electrodes, is a salt solution. Most likely you will be using a solution of copper sulfate ($CuSO_4$) or zinc sulfate ($ZnSO_4$). This is an electrolytic cell -- one which requires a current input -- as opposed to a galvanic cell -- one which spontaneously produces a current (a battery). The circuit is shown in Figure 2-8.

> **Question 1:** Which electrode will be positively charged and which negatively charged when the circuit is closed and a potential applied? What is the charge carrier in the solution? What is the charge carrier in the wire? Does the solution develop a charge?

> **Question 2:** Postulate: What will happen at the cathode? If you expect something to grow, sketch what you expect it to look like as it grows.

Draw three stages of its development: early, half-way through, and completed. Explain why you expect it to appear the way you have drawn it. *Save your sketches in your laboratory notebook.* You will want to compare these predictions with later results.

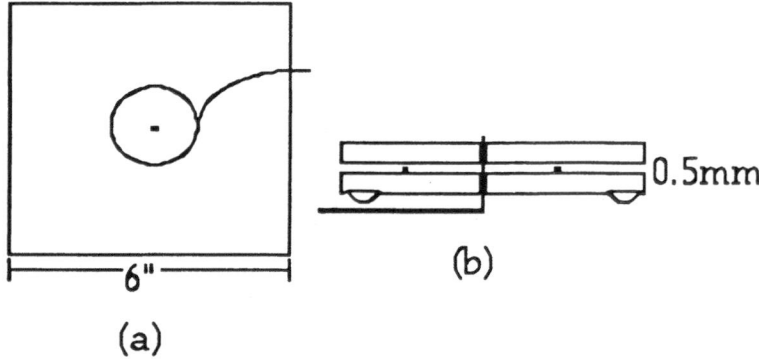

Figure 2-7. The electrodeposition cell. (a) Top view. Circular loop of wire is the anode, surrounding the central cathode. The space between anode and cathode is filled with electrolytic solution. (b) Side view. Spacing between plates is thickness of the anode wire loop. Cathode wire extends vertically through both plates.

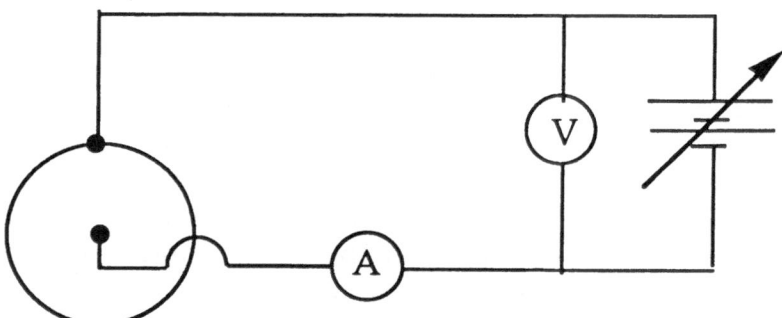

Figure 2-8. Circuit diagram for electrodeposition experiment. The cell schematic is in the lower left-hand corner. The meters labeled A and V measure the current and voltage, respectively. An adjustable voltage power supply is indicated.

2.4.3 Brief Theory Review:

For what follows you will need to recall or understand that:

a) In the electrochemical deposition (ECD) experiment, atoms are *reduced* at the cathode surface. A copper ion (for example) gains an electron and precipitates out as a copper atom, so that the cathode grows.

b) Avogadro's number is 6.02×10^{23} molecules = 1 mole of molecules.

c) The concentration of a solution is measured in molarity (symbol M). The molarity refers to how many moles of solute there are in a liter of solution. For the purposes of these experiments, we will *not* take into account volume expansion when the solute is added to water. So, we will interpret 0.1 M $CuSO_4$(aq) as being 0.1 moles of $CuSO_4$ in 1 liter of water. (The abbreviation "aq" stands for the Latin "aqua," which means water—and is the root of the word "aquarium." We speak of an "aqueous solution.")

d) A coulomb is a measure of electric charge which is equal to 6.24×10^{18} elementary charges (for example, the charge on the electron). An ampere is a measure of current: 1 ampere = 1 coulomb/second.

e) It will aid your reasoning about the experimental behavior to recall the relationship between voltage V and electric field E. If a voltage V is applied between two large plates separated by a distance d in vacuum then the magnitude of the electric field between the plates is: $E = V/d$. The electric field points away from the positive plate, and towards the negative plate. A positive charge moves in the direction of the electric field. For the geometry of the electrolytic cell used in this experiment (a single-post cathode circled by an anode) the relationship is more complex; however the basic proportionality is roughly maintained.

f) In thinking about the flow of ions in the electrolytic cell, you should take into account the resistance of the salt solution. Recall that the resistance R in ohms of an object of cross-section A, length L, and **resistivity** ρ is: $R = \rho L/A$.

g) The applied voltage V, resistance of the circuit R, and the current which flows I, are related by the equation $V = I R$. For example, if the applied voltage across your electrolytic cell (i.e., between the anode and cathode) is 10 V, and the resistance of the cell 100 Ohms, then the current which flows is 0.1 Amps.

Mental lubrication: a) A piece of copper wire is 1 cm in length, has a cross-section 1 mm, and resistivity $\rho = 1.67 \times 10^{-6}$ ohm centimeters. What is the resistance of this wire? If a voltage $V = 10$ volts is applied between the ends of the wire, what is the electric field along the wire? What current flows in the wire? What is actually moving in the wire?

b) Now think of a tube of 0.2 M $CuSO_4$ solution of the same dimensions as the wire in part (a). The resistivity of the solution is $\rho = 0.77$ ohm-cm. Answer all the same questions as in part (a). Also, can this current flow be maintained indefinitely?

2.4.4 Experimental Procedure

You will need at least one partner, and preferably two partners, for this experiment. Equipment at your laboratory bench includes:

1) the wiring, meters, and power supply for Figure 2-8.

2) the plastic plates of the electrochemical deposition cell

3) electrolyte solutions

4) a ruler

5) a stopwatch (or a large wall clock with a second sweep that everyone can read)

2.4.5 Experimental Preliminaries:

Question 1:   What is the carrier of current in the wires of the circuit? What carries the current in the electrolytic cell? When the cell is hooked up, will current flow as long as the power supply stays turned on?

Question 2:   How can you determine the number of atoms being reduced at the cathode per second? The data can be obtained using the meters in the circuit of Figure 2-8.

Question 3:   Without using a balance, how can you determine experimentally the mass deposited per second? Is there a way to use the ammeter to do this? Is there a way that uses a balance? Do you expect the mass deposited per second to vary with time?

2.4.6 Doing the experiment:

1) Divide the measurement tasks with your partner. You will need to measure and record the following during the course of your experiment:

> a) the time at intervals of 20 seconds (longer or shorter depending on the way you assemble the cell; start with 20 seconds and after you have run the experiment reason whether you should use longer or shorter intervals);
> b) the radius of the deposit at each time interval;
> c) the current at each time interval.

With respect to (b), the growth may not be symmetric and some branches may stop growing during the experiment. The investigator recording the "radius" may choose to measure an "average" branch to characterize the size of the aggregate.

A typical procedure might be that one student watches the clock and announces "Mark" at each interval. At this point the student reading the ammeter reads off the current, and the student measuring the aggregate reads the radius of the deposit. The student who is timekeeper records these two numbers.

2) Hook up the circuit shown in Figure 2-8. The exact shape of your anode is not critical, just be prepared to interpret your results in keeping with the shape you choose. Also, try to keep the diameter (or effective diameter if not a circle) of your anode to between 2 and 4 cm. Follow your teacher's instructions with respect to turning on your power supply. Do not turn it on until instructed to do so. (Different classes will use different kinds of power supplies. Each has its own characteristics. Your teacher will tell you how best to use the power supply in your class.) Depending on the time available, you may want to try several experiments with different applied voltages. As a guide, 1 and 20 volts are probably the low and high values. Try working between 10 and 15 volts to begin with.

3) Grow your aggregate. Take care not to let the aggregate grow so large that it reaches the anode. What might happen if it does?

4) Study your experimental result. Can you explain: asymmetries? branches if your aggregate has them? size of the branches? color?

5) Follow instructions in Appendix C or D for measuring the fractal dimension, depending on whether you are using either a scanner or video camera. At the end of each of these appendices are instructions on how to measure the fractal dimension of the aggregate using an Apple Scanner. Since your aggregate is fragile, proceed to this as soon as you finish your experiment.

2.4.7 Data Analysis for Electrochemical Deposition

It is likely that the aggregate you grow does not appear to be a solid disk. We want to measure the fractal dimension D of this aggregate. To estimate this dimension, we will use, in effect, the circle method, described in Section 2.3. For the radius r of different circles, we will use the approximate radius of the aggregate at different times during its growth. Instead of counting boxes inside a circle, we will (indirectly) count the number N of copper atoms. (The following analysis parallels that in Appendix A.) If the deposit is a fractal, we expect that the number of copper atoms N within a radius r is given by:

$$N = c\,r^{D} \tag{1}$$

Take the log of both sides of this equation:

$$\log N = \log\!\left(c\,r^{D}\right) = D\log r + \log c \tag{2}$$

To make use of this equation to find the dimension D, you need to determine the radius r at a sequence of times during the growth and the number N of copper atoms at these times. The radius you measured directly at several times during the experiment? But what about number N of copper atoms?

Before going further, discuss with your partners how you might measure the number of copper atoms that have been deposited at any time.

1) To find the number ΔN of ions deposited in each time interval Δt, you can multiply the electric current during that

time interval by the length of that interval. That is, $\Delta N = I\, \Delta t$. The units are (coulomb/second) second = coulomb. If you are depositing Cu ions, is ΔN the number of ions deposited per second?

You probably noticed that the current varies over each time interval. Which is the "right" value of the current to use for a given interval? The value at the beginning? the value at the end? Would it make sense to use the average?

2) Since you can compute ΔN for each time interval, you can do a running sum and find the total number of ions deposited after any total time t. This is N(t), the total number of ions deposited as a function of time. But you can also measure the radius r at time t. It follows that you know N(r), the total number of ions deposited as a function of r.

On LogLog paper we automatically plot the quantity log N versus the quantity log r. It is equivalent to setting $y = N(r)$, $x = \log r$ and $b = \log c$ and plotting $y = D\, x + b$. If you didn't have LogLog paper, you could use ordinary linear graph paper and press the logarithm button on your calculator to make exactly this same plot. *If the resulting data appears to lie on a straight line*, then D, the fractal dimension, is the slope of that line.

3) Using LogLog graph paper or a calculator and linear graph paper or Cricket Graph or similar plotting program, plot log N versus log r. *If the data appears to lie on a straight line*, measure the slope D, your result for the dimension of the growing aggregate.

4) What is the total mass of the aggregate you deposited?

5) In the cell geometry you are working with, the typical thickness of the deposit is about 60 microns. The density of Cu metal is 8.92 gm/cm^3. What would be the radius of a solid copper disk with the same mass as your final deposit?

2.5 Computer Measurement of Fractal Dimension

Activity 2.9 Fractal Dimension Computer Program

We learned with the **Coastline** program that a computer can quickly cover a pattern with boxes of various sizes, count the number of boxes, and plot the result in order to measure the dimension of that pattern. In this section we learn to use a new and even more powerful computer program called **Fractal Dimension** to analyze patterns grown in the laboratory, as well as patterns grown by computer programs such as **Coastline**.

2.5.1 Remeasure Coastline

1. Open the **Fractal Dimension** program.

2. Pull down the **Images** menu and import the Coastline image you saved earlier as a MacPaint file.

3. At the upper left of the screen is a little box with overlapping circle and box icons. Click so that the box icon comes to the front and is colored red. This activates the box or grid method for measuring dimension.

4. Move the **Length** slider at the top of the screen to give you any desired box length, then click on the **Count** button to have the machine cover the coastline with boxes.

5. Change the length setting and repeat with several sizes of boxes.

6. At the left is a second window containing a LogLog plot. Bring this window to the front, look at it, and fill in gaps in the plot by covering with boxes of appropriate size. This plot gives the value of the slope, which is equal to the dimension.

7. A third and final window is a data table listing box lengths and counts. If you feel that one or more of the data points on the graph should be eliminated from the slope measurement, click on that row of the data table. This will "gray out" that row and the dot for that entry will disappear from the graph. (Restore the data point by clicking again on that item in the table.)

8. Compare the dimension you measure for your Coastline with that measured in the Coastline program itself.

2.5.2 Measuring other Patterns with the Fractal Dimension Program

Analyze other images from the **Images** menu using the box method. In particular, you may want to check that a straight line (whether horizontal or diagonal) has a dimension of one. Is a solid square 2-dimensional, according to the **Fractal Dimension** program?

2.5.3 Measuring the Pattern from the Electrodeposition Experiment

Now call up the image of the pattern you grew in the electrodeposition experiment and measure its dimension using the box method. How does this value compare with the value of the dimension you measured using the current and radius during the experiment itself?

2.6 Modeling the Growth of a Fractal Pattern

Activity 2.10 Building an Aggregate by Hand

Purpose: To carry out a hands-on activity that models the aggregation process.

How can we describe the process by which a pattern grows (aggregates)? Can we mimic the way a charged atom (ion) in a solution dances around, then plates out (becomes an uncharged metal atom)?

What does it mean for an ion "to dance"? Dancing means to stagger around randomly. A dancing ion is taking a random walk! We can use our understanding of random walks to mimic the process of electrochemical deposition.

Another word for *mimic* is *model*. We *model* the aggregation process using our knowledge of a random walker who staggers around and sticks to a growing structure. In the following activity you will use a 2-dimensional random walk to mimic (model) the aggregation process.

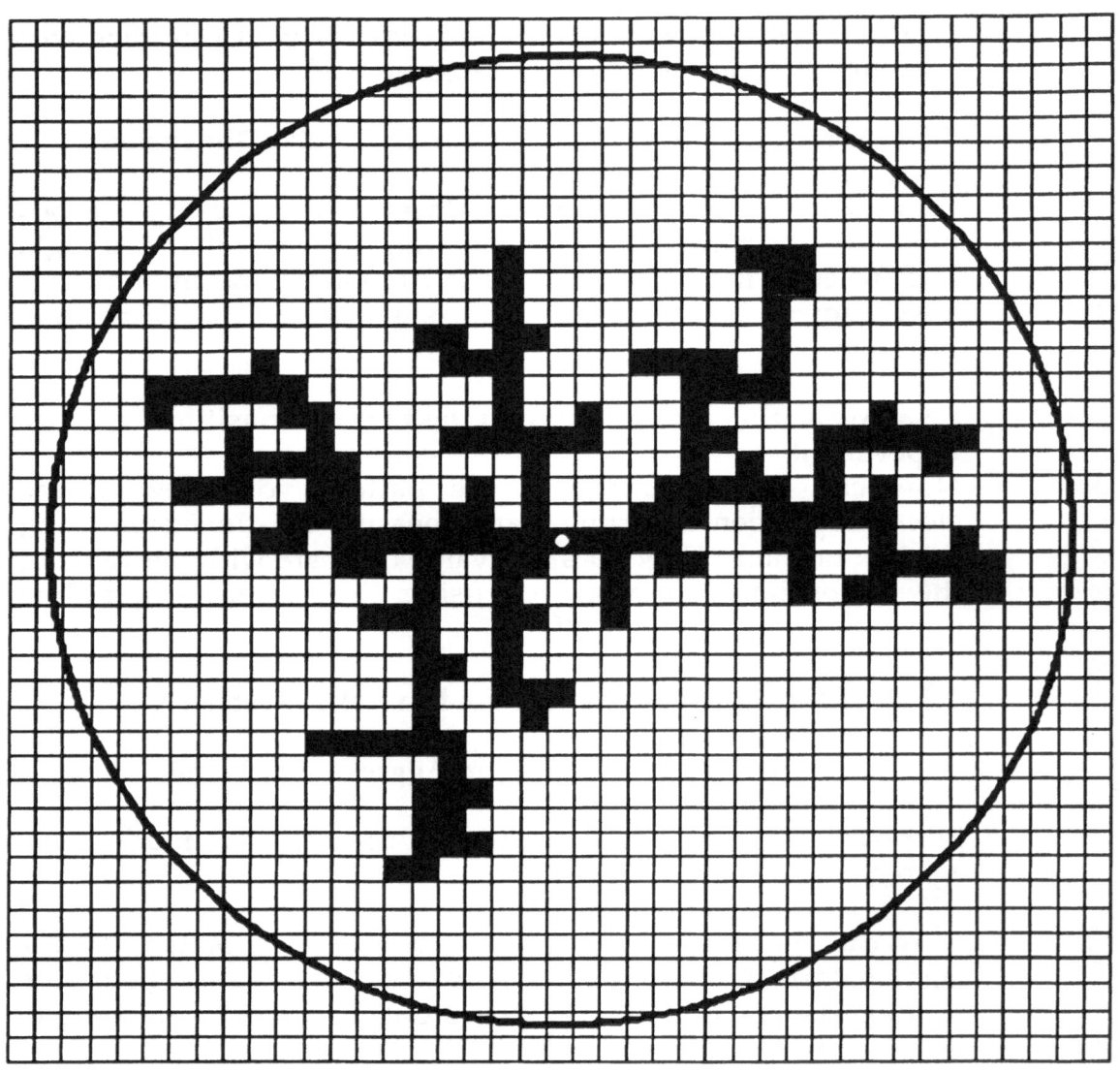

Figure 2-9 An aggregation resulting from a random-walk
model. This pattern was built from a single "seed"
(black square with white dot) by releasing 170 random
walkers one at a time from the rim of the surrounding
circle and allowing each to walk until it reached and
stuck to the growing pattern.

Figure 2-9 shows a small random aggregation pattern. This
pattern was built using the same method you are about to use.
Your job is to start a random walker near the pattern, on the rim
of the surrounding circle. Then let the walker stagger around
the grid until it reaches the black pattern. When the walker
reaches the black pattern, it sticks to the pattern and becomes a

black square. Then another walker starts from a random point on the rim of the surrounding circle and staggers around until it sticks to the pattern. And so forth. That's all there is to our model except for one minor detail: If a walker steps outward across the circle, it is removed and a new one started from a random location on the circle.

Do you think that the electrodeposition experiment can be described using this model?

Activity: You will be given a copy of Figure 2-9 on which to draw. Working in pairs, do the following:

0. Choose a small object to stand for your walker: something small enough to fit on one square; something that will not roll.

1. Choose a starting place for the walker on the rim of the circle by spinning your pencil like a propeller as you drop it onto the figure. Place your walker wherever the tip of the pencil points on the rim of the circle.

2. Now roll a 4-sided "bone" and move your walker one space right, left, up, or down according to the following scheme:

   - If the bone comes up with the number 1 -- move one space right
   - If the bone comes up with the number 2 -- move one space left
   - If the bone comes up with the number 3 -- move one space up
   - If the bone comes up with the number 4 -- move one space down

(You can use a six-sided die; when you get a 5 or a 6, do not move the walker, just roll again. Some game stores sell tetrahedral—4-sided—dice.)

There is one exception to the above rule: If the next move takes the walker outside the circle, remove the walker and start a new one from another random location on the edge. This keeps the walker from wandering away from the pattern.

3. Repeat rolling the bone and moving the walker until your walker lands next to a filled square of the pattern: right, left,

above, or below the filled square (NOT a diagonal position). When this happens, fill in the square where the walker is sitting. That is the new addition to the aggregation.

> It may take quite a while to add one more filled square to the aggregation. You spend a lot of time rolling and reading the bone. To shorten this time, we have programmed a computer to roll the bone and print out a table, Figure 2-10. In this table L and R mean "left" and "right," respectively, while U and D mean "up" and "down." To read the table, start anywhere you want and read in a straight line horizontally or vertically, without repeating any line.

4. Spin the pencil again to point to a new starting point on the circle. This time one of you read out loud the moves from Figure 2-10 ("up," "right," "up," "down," etc.) while the other one moves the walker. How fast can you make your moves using the new method? Remember, whenever a walker moves outside the circle, remove it and start a new walker.

5. When the walker lands next to a filled square of the aggregation (to the right, left, above, or below), fill in the square and start again.

```
L U L U R U R R R U D D U D U U L R U L R R D L R
R D R R D L L D U D L U D U D U D R L L D D R U R
L R L D D R R D L R D D D D R U D U L R D R D U L
D U L L D R L R U U R R U D L U U D L L R R R D U
R L R D L R D U D D R U L U L D D R L R R U D D D
L R D L D R D U R L U D D L L D R R L U U L U R U
U R U U R U U R L D R D U L U D U D U D D U R D D
D R L L U U U L R L U U U L U R U R L L L U D L D
R L L R D R U R R U U U U U U D R L L U U U U L D
R U D U L R R U U D L D D U L L L L R L R D L D L
R D L L D R U U L U U U L D L U D D L L D D R U L
D R R L L L D D R D R L D R R D R R U D D D R L
U D L R U U L R D U D U R L D D D R D R U D U D R
D R R R L D R R U R R L D D U U U R R L R U L D D
R D R R D U R D R U U R U L U U D D R L D D L U D
D L R L R D R L L R L R L R L L R R R R L D D D L
L D D D U L D L D R U L U R D U L L D R D R L L U
R U D D D L L R R D L L L R U R L U L L U R U U L
U U U U D R U D L D D R D R R L U U L D R D L R L
D R L R U R R U L R L L L D U U L L R L U L R D R
U L U L L R U R L D L D U R U D R L D R D L L U R
R L D U L D L R D L U U D U U U U R D R L U L R D
L D D D D U R U L L L L D L L R R U L D R D L L L
R U U U U D R U D L U U D D D R R L D L U U L L L
U D D L L L R D R R U R R U D L U D D L U U D R U
```

Figure 2-10 Randomly-generated table of steps up, down, left, right (U, D, L, R). Read a line of letters from the table to direct the steps of your 2-D walker.

6. Continue this process until you have added about 10 filled squares to the aggregation.

7. Compare your results to those of others in the class. Are the new squares mostly added near the ends of the spidery legs or near the center of the existing pattern? Why should this be? Can you find a simple explanation?

The process we are modeling has the scientific name *diffusion-limited aggregation*, or *DLA* for short. The name comes from the fact that the growth rate of the pattern depends on the rate at which particles arrive at the surface by diffusion.

Compare this model to the dance of ions in the electrodeposition experiment. What component of the experiment does the walker represent? What component does the black pattern represent? Who or what is doing the "bone throwing" in the electrodeposition experiment? In what ways do the ions in the electrodeposition experiment move differently than the walker in the aggregation model?

Activity 2.11 Building an Aggregate by Computer

Can this process be made faster? Of course! Let the computer make random choices *and* move the walker. The computer moves the walker much faster than we can; still, it uses the same simplified, stripped-down mechanism to mimic a complicated process in Nature.

Open the **Aggregation Kit** program. On the right is a blank screen with a so-called "seed" molecule in the center.

1. The settings on the left regulate the process by which the seed will grow. For now, make these settings:

- **Upper slider:** Speed. Set to a position near the left: 10 steps/sec.

- **Bottom slider:** Magnification:. Set to a middle position.

- **Bottom right buttons:** Particle movement. Select **Random**.

2. Press **GO** (upper left button).

A random-walking "molecule" is released far away from the seed and staggers back and forth, up and down, until it finally touches and sticks to the seed. Immediately a second random walker is released, ultimately striking and sticking to the pair at the center. Then the third, fourth, fifth, ..., hundredth walker adds to the growing ungainly pattern.

> In order to speed up the process, if the walker is too far from the aggregate, the computer does not continue the random walk step by step. Instead, it places the particle at a point nearer the seed, in a manner consistent with the particle reaching this nearer point by a random walk.

3. Now try new settings on the sliders: faster speed, smaller cell size.

> Why is the DLA pattern so "leggy," so "spidery"? Why isn't it a compact blob? In your answer, think of the growth process in detail: The walker staggering back and forth as it passes inward between the ends of two legs.

4. Can you find settings for which the resulting pattern looks like the result of your electrodeposition experiment? When you develop such a pattern, choose **Save as MacPaint** under the **File** menu, using a name you will remember. You will need this saved pattern later.

What Do You Think?

Get together with your lab partners and try to answer the following questions. Be prepared to share your opinions with the rest of the class.

(1) You have just discovered how a random walker can be used to build pictures of spidery patterns found in nature. Does this mean that random processes are actually involved in forming these objects in nature? Or are lightning, nerve cells, termite tunnels, and electrodeposition patterns formed by totally different processes, processes that have nothing to do with random walkers?

(2) The discovery that a random walker can be used to build pictures of the spidery patterns found in nature is very important to scientists. Why do you think this discovery is important? In what ways could this discovery prove to be useful?

Changing Particle Movement.
1. Start a new pattern by choosing **Clear** from the **Edit** menu, then click on **Split**. This creates a new window showing a duplicate of the initial seed. For this second pattern, click on the **Straight** button under **Particle Movement**.

2. Click **Go** to start movement in this second window.

3. Now click on the first window and click **Go**. Click on one window then the other, changing the speed of each so that the two patterns grow at approximately the same rate. As each patterns grows, keep it all on screen by moving the **Cell Size** slider to the left.

- How is the movement of the molecule different for the **Straight** setting than for the **Random** setting?

- Does the pattern grow in more steps or fewer steps for the **Straight** setting than for the **Random** setting?

- Is the pattern more leggy or more compact for the **Straight** setting than for the **Random** setting?

4. Using the **Straight** setting, can you make a pattern that is similar to the result of your electrodeposition experiment? If so, save it as a MacPaint file for later analysis.

Measuring the computer-generated aggregation model

Close the **Aggregation Kit** program and call up the **Fractal Dimension** program. Import the pattern you saved from the **Aggregation Kit** and measure its dimension using the box method. If you saved an image from the **Straight** setting of walker motion, call that up and measure its dimension also.

The **Fractal Dimension** program offers a second way to measure the dimension of an object, the so-called circle method (Section B above). Try out this second method on your computer-generated aggregate.

1. Click on the box/circle icon at the upper left of the screen, bringing the red circle to the front.

2. In the center of the screen is a kind of cross-figure that indicates the center of the circles you are about to have the computer draw. Drag this cross-figure to the approximate center of your aggregate image.

3. Now select a radius for the first circle using the slider at the top of the screen and click on the **Count** button. The computer counts the number of screen pixels inside the circle of that radius.

4. Select several more radii and have the computer count for these circles also.

5. Call up the graph and the data table and fill in empty portions of the line. Click on data table entries to discard points you do not want included in the graphical measurement of slope.

6. Do you obtain the same measure of dimension from the circle method as from the box method?

7. You may want to use the circle method to re-compute the dimension of the saved computer coastline or the saved image from the electrodeposition experiment, comparing these measures of dimension from those obtained from the box method.

8. Which setting of the **Aggregation Kit** program can produce a pattern with a fractal dimension nearer to that of the pattern from the electrodeposition experiment: the **Straight** setting or the **Random** setting for the way the walker moves?

9. There is always *some* difference between the dimension of the pattern grown in the electrodeposition experiment and the pattern grown with the **Aggregation Kit**. What does this difference mean? How large a difference is acceptable if the **Aggregation Kit** model of fractal growth is to be accepted?

See Section 2.11 for suggested research projects on diffusion limited aggregation.

2.7 The Hele-Shaw Experiment

2.7.1 Introduction

How is the recovery of oil from old oil fields related to the development of ulcers? Is the path of a lightning bolt governed by the same basic physics as the growth of a snowflake? And why can you understand the utility of a lightning rod by studying random walkers? or by studying the growth of a dust particle?

The experiment described in this section exhibits aspects of the fundamental physics of all the processes described above. This experiment, a study of "viscous fingering" was originally performed by Mr. H. S. Hele-Shaw, a naval architect, in 1898. His chosen geometry was a little different from the one we use here. Nevertheless, we refer to the apparatus as the **Hele-Shaw cell** after his original design.

2.7.2 Brief Theory Review

Two little stories:

The Birthday Party

It's your friend's birthday and you are throwing a surprise party. Time to blow up balloons. You do not have a balloon pump, so you do it using lung power. What kinds of balloons are hardest to inflate? Are long "dachshund" balloons harder or easier to inflate than spherical balloons? And when is any balloon hardest to blow up? Is it harder to blow air into a balloon when it is deflated? Or when it is close to full inflation?

What is the force that resists as a balloon is inflated? After the balloon is inflated, is the air under greater pressure in the interior of a balloon or outside the balloon? If you think it is greater inside, what counter-force keeps the skin of the balloon from exploding under this unequal outward and inward force? And how is this problem related to stretching a rubber band, anyway?

Surface tension is a very common force which shapes much of nature around us. In effect, it is a force which resists the creation of surface area. Thus, when you fill a glass to its lip, and keep filling it some more, the curved surface (meniscus) that allows you to fill the glass beyond its top is maintained by surface tension. When it suddenly breaks, the flowing liquid has greater surface area. Similarly, the water that comes flowing out of your faucet forms a stream rather than a broad band spray because of surface tension -- surface tension keeps the water stream together. Surface tension keeps a balloon from looking like a porcupine when you blow it up.

Can you describe what might create surface tension at a molecular level? How do you envision the molecules at a liquid surface behaving? What about in the case of the balloon? Write a short paragraph describing and contrasting a liquid surface with the surface of a balloon.

Drinking Through a Straw

Think of a straw with water flowing through it. Where is the water's speed largest? At the inner surface of the straw? Or in the middle of the straw? If I wrap the straw with a blanket of ice, and then surround the ice itself with as perfect a thermal insulator as possible, will the water flowing inside the straw eventually cause the ice to melt? Or will the ice remain frozen?

It has been observed empirically that for fluids flowing past a surface, the speed of the fluid at the surface itself is zero. But at the center of the stream the fluid certainly is moving! How can this be? What happens when you move one surface over another? Is it easy? What will stop a metal block from sliding over a metal surface? If you think it is friction, how do you imagine friction operating at a molecular level? Is heat generated? If so, where does that heat go? Can you generalize this argument to the fluid flowing in a straw, or over a plane surface? What is happening at the molecular level that causes adjacent layers of fluid to move at different speeds?

The frictional property of fluids is referred to as **viscosity.** The greater the viscosity of a fluid, the greater the force necessary to maintain such a fluid flowing in a straw, or over a surface. If two fluids have different viscosities, can you imagine what differences in the molecules of the two different fluids could give rise to the difference in viscosity?

The viscosity of various liquids plays an important part in Hele-Shaw experiments. Viscosity is a measure of the resistance of a liquid to flow. At room temperature, honey or molasses do not flow easily: they have high viscosity. Motor oil flows more easily than honey; it has a lower viscosity. Water has a still lower viscosity. Air flows so easily that one might be tempted to say that it has zero viscosity. But it does resist flow a little, as you can prove by trying to breathe through a drinking straw. The viscosity of water is approximately 50 times greater than the viscosity of air

---

Looking Up Viscosity

Pour a puddle of honey on the table. Place the flat bottom of a dish on the puddle of honey. Now drag the dish sideways along the surface of the table. It takes a force to keep the dish moving. The viscosity of the honey is related to the force needed to drag the dish over the honey-puddle. If you carry out the Hele-Shaw experiment using other fluids, you may want to look up their viscosities. Viscosity is measured in poise, whose plural is also poise. The viscosity of water is easy to remember; it is one centipoise, that is, one hundredth of a poise. Glycerol has a viscosity 1200 to 1400 times greater than water, or 12 to 14 poise. Air has a viscosity some 50 times smaller than that of water, or approximately 200 micropoise.

Activity 2.12 The Hele-Shaw Experiment

2.7.3 The Hele-Shaw Equipment

Figure 2-11 shows a sketch of the Hele-Shaw cell. The two plates of plastic are separated by cover slips (little squares of glass or plastic usually used to cover microscope slides). In each corner there are two cover slips, one on top of the other. Typically glass cover slips are between 130 and 160 microns in thickness. Plastic cover slips are roughly the same thickness. So the spacing between the plates is roughly 300 microns.

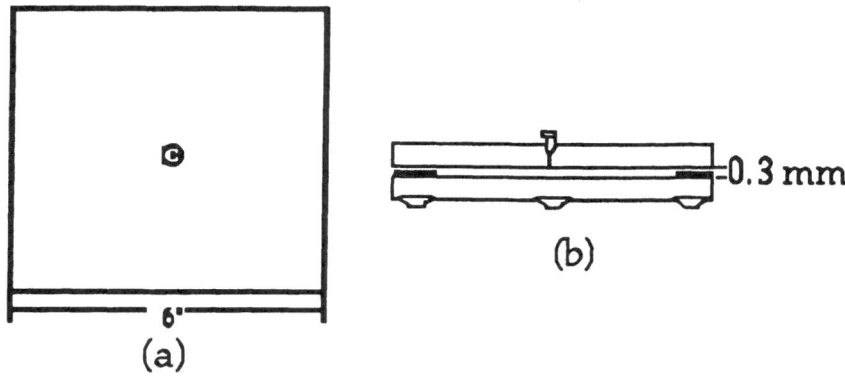

Figure 2-11. The Hele-Shaw cell. (a) Top view. Fluid is injected through the center hole. (b) Side view. Spacing between the plates is determined by the thickness of two cover slips, one on top of the other placed in each corner of the cell

In the experiment you will first inject glycerol through the central nozzle to fill the cell. After the cell is filled with glycerol, you will inject colored water.

2.7.4 Speculate and Hypothesize

Before carrying out the experiment, answer the following questions.

Question 1: You will inject glycerol into the empty cell. As the interface of glycerol into air advances, what will its shape be? Sketch your predicted shape, and provide a written argument to support your sketch. Was your conclusion influenced by: the viscosity difference between the glycerol and air? symmetry of the cell? narrowness of the cell spacing? surface tension between the glycerol and air? prior experience?

Question 2: The cell is filled with glycerol. With another syringe, you now inject water through the central nozzle. The water has food color in it so that you can distinguish it from the glycerol and spot the interface between the water and glycerol. As the water advances, what is the shape of the interface between the water and the glycerol? Sketch your predicted shape and write a brief argument supporting your sketch. Does your argument depend on the viscosity contrast between the water and glycerol? symmetry of the cell? narrowness of the cell spacing? surface tension between the water and glycerol? prior experience?

2.7.5 Experimental Procedure

Hele-Shaw with Glycerol

Using the apparatus sketched in Figure 2-11 you can now perform the experiments suggested above. For data collection it will be easiest to work with one or two partners. If it is possible, use a video camera and either videotape the experiment as it progresses, or capture the experiment to a computer.

1) Assemble the cell as indicated in the figure. Fill a syringe with approximately 20 cc of glycerol. Attach the syringe to the center nozzle, and inject the glycerol. If you are very patient, you can let the glycerol flow by itself into the cell. Otherwise, you can drive the plunger of the syringe slowly.

As the glycerol enters the cell, study its interface carefully. If bumps develop on the interface, watch what happens to them. By a bump we mean a deviation from a symmetrical, smooth interface. If you see a bump, does it grow? Or does it recede with time and the interface "heal" itself? What might give rise to the development of these bumps? What determines their lifetime?

As the glycerol approaches the edge of the cell what is the symmetry of the interface?

If a video camera is available, record the following injection process.

2) The central nozzle on the plate should be filled with glycerol to the top. Fill the 1 cc syringe with water containing food color. Carefully attach the syringe to the nozzle. Inject 0.2 cc of the glycerol (at a rate of, say, 0.1 cc per second), watching the interface *carefully*. Measure and record the radius of the interface. If it is not symmetrical, approximate the average radius. With injections of 0.2 cc at a time, repeat this measurement process until all the water has been injected.

3) Scan the interface with AppleScan, or grab the image of the interface using VideoSpigot and a video film.

Hele-Shaw with Carrageenan

The purpose of this experiment is to observe the viscous fingers which form when fluids of different viscosity interact with each other, *and when there is a chemical reaction due to differences in the pH of the two fluids.* Specifically we will be working with carrageenan of varying concentrations and pH. Carrageenan is a polymer which in aqueous solution gels at high pH (basic conditions), but does not gel at low pH (acidic conditions). Carrageenan is added to ice cream, whipping cream, and some other foods precisely because of its gelling properties.

The viscosity of aqueous carrageenan solutions is strongly dependent on both concentration and pH. In the previous experiments in the Hele-Shaw cell you worked with glycerol. The viscosity of glycerol is a simple function of its concentration. By contrast, for carrageenan, the viscosity does not depend simply on the concentration. This is because the long chain molecule of the carrageenan polymer tends to intertwine with other carrageenan molecules. The result is a strong dependence of flow properties, such as viscosity, on concentration and flow velocity.

pH is an important parameter in carrageenan solutions. Long-chain molecules such as carrageenan can interact through electrostatic forces. The pH of the solution can strongly influence this interaction. The pH reflects the ions present in the solution, which mediate this interaction.

For the experiments described below, we recommend working with four different carrageenan solutions: concentrations of 5 mg/ml and 10 mg/ml, each with a pH of 3 and 5. We encourage you to try other concentrations and pH. These four, however, will produce representative results. Prepare your carrageenan solutions according to the instructions provided in Appendix E.

1) See step 1 of the Hele-Shaw experiment with glycerol.

2) See step 2 of the Hele-Shaw experiment with glycerol.

2.7.6 Data Analysis for the Hele-Shaw Experiment

Theoretical Review:

The reason that fluids flow is because of pressure differences across the fluid. For example, when you suck on a straw the fluid flows up the straw because atmospheric pressure at the bottom of the straw is greater than the pressure in your mouth. Similarly, when you apply pressure with the syringe at the central nozzle of the Hele-Shaw, fluid flows in the cell because the pressure at the center is greater than the atmospheric pressure surrounding the open edges of the cell.

To understand what follows it is useful to know:

a) Usually the speed of fluid flow in a pipe, or between two plates, is proportional to the pressure difference across the fluid, and inversely proportional to the distance over which that pressure difference is maintained. For example, if you suck harder on a straw, reducing the pressure in your mouth more, this increases the pressure difference across the straw (since atmospheric pressure remains constant), and the fluid flows faster. Similarly, if you use a shorter straw, the fluid also flows faster, since the same pressure difference exists across a shorter length. In the case of the Hele-Shaw cell, if your plates had a shorter edge, then for the same applied pressure on the syringe, the fluid between the plates would flow faster.

Question: Can you draw an analogy between the applied pressure in the Hele-Shaw experiment and the applied voltage in the electrochemical deposition experiment?

b) For a given applied pressure difference, the speed of fluid flow between the plates of the Hele-Shaw cell is proportional to the square of the spacing between the plates. Thus, if you double the spacing between the plates, you quadruple the flow rate if all other conditions are kept constant.

Observational Analysis:

1) When you injected the glycerol or carrageenan into the air-filled cell, you watched the interface for bumps. When a more viscous fluid invades a less viscous one, e.g., the glycerol or carrageenan invading air, the interface is "stable" in the sense

that any bumps which form die away. The interface "heals" and returns to being smooth and symmetric. If you videotaped your experiment, or recorded it into a computer, play it back and watch this phenomenon again. Alternatively, repeat the experiment if you did not observe this phenomenon the first time.

2) By contrast, when you injected the water into the glycerol, or the acid into the carrageenan, the interface did not remain symmetric but broke up into **viscous fingers**. Viscous fingers are what develop when bumps on an interface grow. We say that the interface is "unstable" when a less viscous fluid is injected into a more viscous one.

Despite the fact that bumps can grow in this latter case, repeat the experiment, or study your film of it, and try to determine *when* the bumps start growing. Is there an initial symmetric interface at the outset? Or do viscous fingers emanate from directly beneath the nozzle?

3) Either by repeating the experiment, or referring to your film, study the growth of individual fingers as they advance. Does a finger retain its shape and grow like an inflated dachshund balloon? Or do fingers break up and form multiple new fingers? Do all fingers grow, or do some stop advancing and become static? Where are the ones which become static located? Where do advancing fingers split?

The growth of the branching tree structure of viscous fingers you are observing is a primary example of how branching structures develop. A bump appears at the interface. The bump grows faster than adjacent areas of the interface and develops into a finger. The finger itself then splits because of the "bumps grow" phenomenon, and forms multiple branches.

Computing the Fractal Dimension from Experimental Data

1) During your experiment, for each volume V of fluid injected you obtained the radius R of the pattern. If the fingering pattern is a fractal then we expect the relationship:

$$V(R) = cR^D \qquad\qquad (1).$$

where c is a constant of proportionality and D is the fractal dimension. If we take the logarithm of both sides of (1) we obtain:

$$\log V(R) = D \log R + \log c \qquad (2).$$

It follows that if we plot log V versus log R and obtain a straight line, then the slope of this line will be D, the fractal dimension, and the intercept will be log c.

Using either Cricket Graph or LogLog graph paper, plot log V versus log R and determine D from the slope.

2) Alternatively, we can determine the fractal dimension of the final pattern by capturing the image using a scanner, or by capturing the image to disk using a video camera if the computer has video capabilities. See Appendices C or D for capturing procedures depending on your equipment. After the image has been converted to a MacPaint file, use the Fractal Dimension program to determine D.

2.7.7. Discussion: Why Do Viscous Fingers Branch?

Why do bumps grow faster than the rest of the interface? The answer to this question is fundamental to understanding why the Hele-Shaw experiment gives rise to a tree-like pattern when a less viscous fluid is injected into a more viscous one. Recall from our brief theory review above that the flow velocity is proportional to the pressure difference and inversely proportional to the distance over which the pressure difference occurs.

Consider a bump on a circular interface as in Figure 2-12. The pressure is the nearly the same at bump interface as on the circular portion of the interface. On the other hand, the top of the bump is closer to the edge of the cell than the rest of the interface. As a result, the bump will move faster. *This is why the interface is unstable, bumps grow, and viscous fingers develop.* In a self-similar fashion, subsequent bumps grow on the fingers themselves causing the fingers to branch. And so, a tree develops.

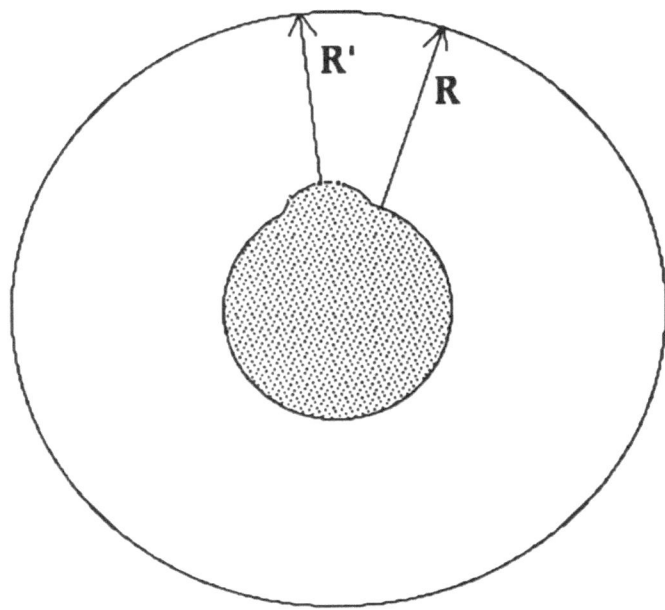

Figure 2-12 Bump on circular interface of injected fluid. R', the distance from the bump to the edge of the cell, is less than R, the distance from the circular interface to the edge of the cell. Because the velocity depends inversely on this distance, the bump grows faster. (The experimental cell described in this text has a square outer boundary, but the argument is effectively the same.)

Question: Can you make an analogy between this process of branching by viscous fingers and the branching process for electrochemical deposits? What corresponds to the pressure in the case of electrochemical deposition?

What determines the thickness of the viscous fingers? Why are the viscous fingers in the carrageenan experiment narrower than those in the glycerol experiment? Predicting the minimum thickness of a viscous finger is still a research issue. A key parameter is the surface tension between the two fluids. Are your results consistent with your understanding of surface tension? Write a short argument as to whether viscous fingers in glycerol are thinner or thicker if you inject air rather than water? Try this experiment.

Write a short argument as to what you would expect for the fractal dimension if you:

a) inject air into glycerol?

b) inject water at higher pressure into glycerol?

c) inject a 50-50 mix of water and glycerol into pure glycerol?

See Section 2.10 for suggested research projects on viscous fingering.

2.8 Branching Patterns Formed by Bacterial Colonies

2.8.1 Introduction

In this section you will study the shapes of bacterial colonies grown under different nutrient conditions. The primary questions are these: Why are the shapes of bacterial colonies grown under starvation conditions similar to fractal patterns of non-living systems, such as electrochemical deposition and viscous fingering. Can we understand these shapes from our experience with random walks and diffusion?

The growth forms of bacterial colonies reflect a current research question: How social are bacteria in colonies? Does the growth pattern of the colony result from purposeful collective behavior by the bacteria? Or is it explainable solely by the physics of multiple random processes, as with a random walker model? A fundamental question is what constitutes social behavior? And are bacteria capable of it? This experiment is interesting, since physicists and biologists are likely to interpret the results from different perspectives. We will ask you to reach your own conclusions as to whether the colony shapes you obtain are determined primarily by simple physics, or whether complex social interactions among the bacteria are necessary.

In this experiment, you will grow a bacterial colony on a flat gelatinous surface containing basic nutrients. Standard laboratory practice is to provide bacteria with a rich source of nutrition. In this experiment, by contrast, the bacterial colonies are grown under varying nutritional conditions which can approach the starvation threshold. This gel will be in a round flat dish (a standard petri dish) about 9 cm in diameter. To start the colony growing, you will put a small drop of bacteria at the center of the round gel. Then you will cover the petri dish and place into an environment (an incubator) maintained at a standard temperature and study the pattern of the colony as it grows outward over a period of days and weeks.

You may wonder: How do the bacteria respond to the condition of low available nutrition? Can they hunt for food? What determines whether they can receive nutrition? Can the bacteria cooperate? What determines the optimal shape of the colony? Is the shape we see the optimal one? And what about the nutrients? Do they behave dynamically? Or are they stationary in the gel waiting for the bacteria to come and graze?

Try answering the following questions: 1) Does the nutrient concentration remain constant in the gel over time? If not, how does it change? Where is it highest? Where is it lowest? 2) Sketch the shape of the colony as it grows. 3) Draw analogies between this experiment and the simulation of DLA growth in the Aggregation Kit.

For many researchers there is a tendency to reason differently about living organisms than about "dead" objects. By now we are well aware that non-living material can be very dynamic. Is there an advantage to reasoning about bacteria by analogy to living organisms as opposed to restricting your analysis to the modes of behavior of non living matter? Try to list the advantages and disadvantages for "projecting" on to bacteria your expectations of living organisms in analyzing their behavior. Does it make sense to ask what bacterial *want* to do? *try* to do? *plan* to do? Do you think that physicists, chemists, biologists, religious people, non religious people, have the same perspective?

2.8.2 Tools for Reasoning about Bacteria, their Nutrition, and Agar

To be able to speculate about this experiment you will need some basic facts about bacteria. When you did the electrodeposition experiment you probably had an internal, mental model for what was happening. You "imagined" copper ions moving through solution. You "pictured" them attaching to the growing aggregate and being reduced (i.e., capturing two electrons). You reasoned about the applied electric field and its ability to drive ions through solution, and electrons through the external circuit. And when you compared your results to those of the Aggregation Kit simulation, you drew analogies between the simulation and the experiment.

In order to build a mental model to analyze the experiment in this chapter, you need certain facts about bacteria and their behavior. You may also find it useful to adopt some of the code words scientists use to discuss these experiments in quick and short sentences. These words include **chemotaxis**, which is shorthand for saying that bacteria can respond to chemicals in their environment by moving; **motile,** which is the biology synonym for mobile; **viscosity**, defined in the Hele-Shaw cell experiment, it refers to the ease with which one fluid can move through a narrow tube or penetrate another liquid; and, **flagellum**, a whip-like organelle which many bacteria use to propel themselves.

Among the bacteria we recommend for this experiment are *Bacillus subtilis,Escherichia coli*, and *Enterobacter aerogenes*. All are rod-like in shape with dimensions of 2 – 3 μm in length and 1 μm in width. They occur commonly, are usually found in the human digestive tract, soil, and dairy products (unless pasteurized!). To grow *B. subtilis* and *E. coli* you will need to incubate the agar plates at 37 C. *E. aerogenes* will grow at room temperature. Typically these bacteria will multiply every 20 minutes if adequate nutrition is available.

Bacteria respond differently under starvation conditions. *B. subtilis* retreats into a dormant mode by forming **endospores**. By contrast, *E. Coli* and *E. aerogenes* die.

The wild strain of each of these bacteria (i.e., the most commonly observed strain in nature) is motile. Each is flagellated, although with varying numbers of flagella. Under a microscope, near the colony boundary, bacteria growing on agar can appear like many, many tadpoles swimming randomly. It is not far-fetched to describe each bacterium as executing a random walk, but one restricted to the colony environs. Bacteria also excrete an extra cellular fluid which can enhance their ability to move on the agar surface.

In our experiments, nutrients are supplied to bacteria by dispersing them in a gelatin-like material made from **agar**. Agar is a long-chain molecule derived from seaweed. When dissolved in water at sufficiently high concentrations, it causes the solution to gel. In this way it is much like gelatin. Although it is a gel, nutrient molecules diffuse through agar nearly as quickly as through water.

It is known that bacteria respond to chemical signals. One famous example is that of slime mold — under starvation conditions the bacteria, initially distributed over space and not in direct contact, release chemicals which result in a chemotactic response: the bacteria come together into a single colony and form a single fruiting body to survive the adverse conditions. (In biology *taxis* refers to motion by an organism in response to a stimulus. **Chemotaxis** is motion in response to a chemical stimulus. **Phototaxis** is motion in response to light. These terms are similar to **chemotropism** and **phototropism**, however a tropic response can include growth as well as motion.)

While it is known that bacteria communicate chemically, it is difficult to isolate the chemicals used, or distinguish purposeful organization resulting from this communication from superficial organization apparent to the human observer. As an example of the distinction here: in electrochemical deposition, beautiful patterns emerge as the ions aggregate; moreover, the atoms "communicate" chemically. Nevertheless, we do not ordinarily attribute to copper atoms the ability to engage in social behavior.

Activity 2.13 Modeling a Rough Surface

As part of your data analysis you will measure the fractal dimension of the surface, or perimeter, of the bacterial colony. The question is whether the colony interface may reflect social interactions, or whether the spread of the colony is effectively random.

A surface which is random can be generated by the following random walk exercise. Take a piece of two-dimensional grid graph paper and draw a horizontal line across the middle. This will be your y = 0 line. Make a mark at the left-most point of the axis. This is your origin (x = 0, y = 0). Flip a coin. Assign heads to +1, and tails to -1 motion, in the y direction. Flip the coin. If it is tails, put a mark at (1, -1); if heads at (1,1). Repeat this process moving your mark over one unit on the x-axis each time (the x-axis is effectively time), and up or down one unit in the y-direction, depending on the result of the coin flip. For example, if you your flips are THHTTTHTHHT then your last marked point would be (11, -1).

Connect the points of your mark's trajectory. The result is a random surface. If you have the Rough Surface Analyzer program available, you can generate a long such interface and measure its fractal dimension. Alternatively, connect your points with dark magic marker, scan it into the computer with a threshold so that the grid vanishes, save it as a MacPaint file, and find the dimension using the Fractal Dimension program.

2.8.3 The Physics of Diffusion
A common model of molecules in gases and liquids envisions particles constantly in motion, dancing around because of thermal agitation. On average in a homogeneous gas or liquid nothing changes because of this motion; equal numbers of molecules move in and out of a given volume (provided that volume is much bigger than an individual molecule). Since the motion is random, like that of a random walker, there is no "current", no net directed flow of molecules in a given direction. This is a picture of a medium which is "at equilibrium."

Now extend this mental model to the case where a concentration difference exists. For example, suppose we have a barrier between two containers of water. The container on the left has salt dissolved in the water; the one on the right is pure distilled water. We now raise this barrier carefully, so as not to induce fluid motion. What happens to the salinity on the two sides of the unified chamber? Is there a net motion of salt molecules from one side to the other? Why? Does the random walk of the molecules in solution contribute to this? Does it make sense to define a current of particles corresponding to this motion? If so, will this current flow indefinitely? Why or why not? If not, will it vary in time? Take a guess: What would this current be proportional to?

When a concentration difference occurs between regions of the same medium, that medium is no longer in equilibrium. The result is a net **diffusion** of particles to the region which is concentration deficient from the region which is concentration rich.

> **Question:** As the bacterial colony grows it eats the nutrients which are local to its interface. This results in a nutrient deficit near the interface. But the agar gel began as a

homogeneous medium with a uniform concentration of nutrient. What will happen to the nutrient? Will the colony be able to continue to feed?

Activity 2.14 Growth of Bacteria under Starvation Conditions

As mentioned above, *B. Subtilis*, *E. Coli*, and *E. Aerogenes* are all commonly occurring bacteria. In fact, we are surrounded by so many different bacteria in our environment that to grow a single strain requires careful attention to maintaining sterile conditions.

With respect to safety, although these bacteria surround us, it is still possible to be poisoned by them. Take standard laboratory precautions and wear gloves when handling the bacteria and agar plates.

1. Follow the instructions in the appendix to make your agar plates.

2. Follow the instructions in the appendix to inoculate and incubate your agar plates.

3. In addition to the single point inoculation experiments described in the appendix, do a competing colony experiment. On each of two plates, inoculate at two points, separated by 0.5 cm on one plate and by 1 cm on the other plate.

4. Measure the growth (diameter) of each bacterial colony daily and record it in your lab book. A chart for this purpose of plate number versus day is useful. <u>N.B.</u>: **It is very important that the parafilm be checked daily for cracks and resealed if necessary. Otherwise, your plate will quickly dry up.**

2.8.4 Data Analysis
1. Follow the instructions in Appendices C or D to grab an image of the colony either with a scanner or a video camera.

2. Using the **Fractal Dimension** program, find the dimension of the colony. How does the dimension vary with nutrient concentration in the plates? Measure both the global dimension of the aggregate (the dimension of the entire colony including

the interior), and the surface fractal dimension (the fractal dimension of the perimeter of the colony).

How does the global dimension compare with that of a diffusion-limited aggregate (DLA)?

How does the surface dimension compare with that expected from a random walk? Compare the dimension obtained with that found in Activity 2.13 above. If the dimension differs from that of a random walker interface, does it behave as you would expect from bacteria which are socially cooperative? Write a short paragraph explaining why, or why not.

3. Plot the radius of the aggregate versus time and find the growth velocity. Is this also a function of nutrient level?

4. Is the growth of the colony limited by diffusion of nutrients? If a nutrient molecule moves a distance r in a time t according to the relationship $r^2 = 4Dt$ where D is approximately equal to 10^{-6} cm^2/sec, on average how far can a nutrient molecule diffuse in an hour? A day? How does that relate to the initial and final growth velocities you observed?

5. For the plate you inoculated at two points close together, how do you interpret your results? Is it consistent with a model in which access to nutrients is controlled by diffusion? Can you simulate this experiment using the Aggregation program?

See Section 2.10 for suggested research projects on bacterial colonies.

2.9 Termite Nesting and Foraging in Two Dimensions

On a hot day, sitting out in the sun, watching the ants,. bees, and beetles, it appears that insects roam about in a random fashion. The trails they leave in the dust are only a poor reflection of what we imagine the interiors of their nests must appear like. Not only ants make nests; many other social insects, do so as well. Nests of such insects, especially the highly organized insects, such as termites, wasps, and bees have been compared to living brains with each member of the nest a neuron. Decisions appear to be made cooperatively, as if in a highly organized society.

The more primitive termites feed directly on the wood in which they nest, while the advanced species nest in soil, and forage for dead wood, grass, seed, and other diffuse sources of cellulose. To reach this food, workers extend galleries through the soil, construct covered trailways over the surface of the ground, or march in columns over exposed odor trails.

Studying such insects in the wild is a task for dedicated entomologists. However, it is possible to get a flavor for the nesting and foraging behavior of such social insects using simple apparatus. Below we describe such an experiment using termites.

In the experiment which follows you can study termites forming trails. The termites are confined to sand between two plates of acrylic plastic. See Figure 2-13. The plates are spaced 1/32" apart corresponding to the widest dimension of a termite's body, its head. You will place about 150 termites at the center of the cell, and from there they will forage in the sand seeking food. There will be no food in the sand (although you are invited to alter the experiment—see research activities).

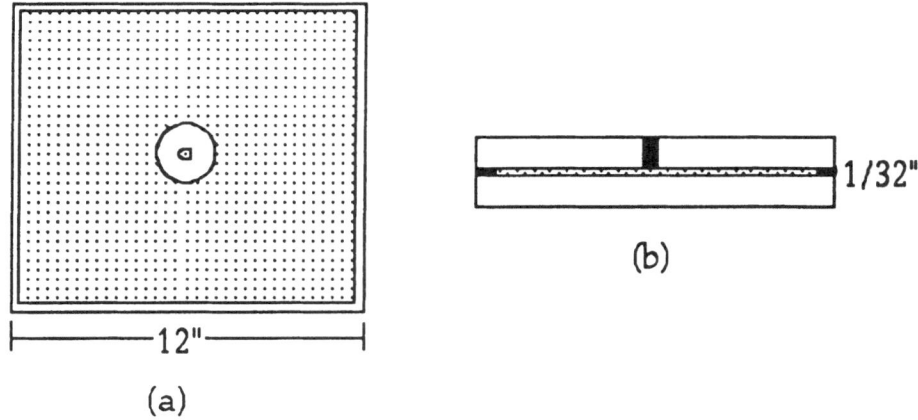

Figure 2-13. The termite cell. (a) Top view of cell. Termites are introduced at the center and move into the shaded area, which is filled with sand. (b) Side view of termite cell, indicating 1/32" spacing between plates.

A fundamental question we will try to answer is whether we can distinguish the termites' foraging behavior from random motion, and whether there is a quantitative way to evaluate the patterns of their trails.

Try predicting the pattern of termite tunnels as the termites spread out from the center of the cell. What would this pattern look like if the termites were randomly walking out from the center?

Activity 2.15 Termite Tunnel Patterns

1.   Follow the Instructions in Appendix H to construct your termite cell, and to load termites into the petri dish on top of the cell.

2.   Keep the cell in a warm but not hot place, and not in the sun. Termites are nearly blind so light shouldn't disturb their activity. However, they dehydrate very easily, and in the cell they could cook.

3.   After about a day most of the termites will have descended into the damp sand. Kick the stragglers over the edge, remove the petri dish, and tape coverslips over the top hole.

4.   The pattern will develop in between one and three days. During this time, take pictures frequently at regular intervals. If possible, use a video camera to grab images. Alternatively, either trace the developing pattern as a function of time, or scan the pattern into the computer.

5.   Watch the termites carefully.   Can you distinguish the different castes and their tasks? How do they communicate? How do they form the tunnels? What is it that keeps the sand from collapsing on them? Do you see any preferred directions to their activity? If so, can you reason why they have selected these directions? Do their tunnels intersect with each other?

Data Analysis

1. Much of your analysis should derive from your observations. Do the termites communicate? Do you think this affects their tunneling? Do you see evidence for trail pheromones?

2. Is the pattern random? The product of a diffusion process? Are the termites effectively ballistic -- do they go in one direction and simply not stop? Is it necessary for them to search every place in the cell?

3. If the pattern development was a diffusive process, what would be the ratio of times to double the effective radius of the foraging? Does this match your observations?

4. Using your scanned or video captured images, measure the fractal dimension of the pattern at successive times. Using the procedure in the appendix, first turn your image into a MacPaint file, clean it up in NIH Image, and then find the dimension with the Fractal Dimension program.

5. Compare your pattern to that obtained with the AntHill Program. The two patterns are different, but can you think of a way that there might be a similarity? Suppose you zoomed way back from your cell, and had many, many more termites, and a long, long time?

See Section 2.10 for suggested research projects on termites.

2.10 Fractal Root Systems

2.10.1 Introduction
The roots of plants are a classic example of branching growth. Every plant has a submerged branching system designed to support its life functions. But why do roots branch? And why do they do so at so many different levels, that is, why are there roots upon roots? Is this the most efficient method for a plant to obtain nutrition? Can you think of reasons for the similarity in the branching pattern of root systems and the patterns controlled by diffusion processes studied in earlier sections?

In the experiments described below you will be encouraged to play detective and try to understand some of the factors which give rise to root structure. In so doing, it will be useful to relate what you learned earlier about diffusion, and the factors which give rise to branching, in inorganic systems. Similar branching structures are observed in snowflakes and during crystallization of minerals. In each case it is a diffusion process that controls the rate at which growth occurs. The branching pattern apparently results from the properties of this diffusion. For crystallization, it is the diffusion of minerals towards the growing surface; for snowflake growth it is the diffusion of the latent heat evolved during crystallization; and for root growth it is presumably the diffusion of nutrients toward the growing roots. This diffusion takes place in solution for hydroponic growth, and in the soil for growth in earth.

One problem botanists face today is the insufficient research done to understand root development and physiology. It is difficult to go out into the field to do experiments, because when plant roots are removed from the soil the roots are necessarily disrupted. One solution to this problem is the one adopted in the experiment proposed below. This is to use a **rhizotron**, a clear-walled chamber through which scientists can observe roots as they grow.

2.10.2 Background

Root growth occurs via elongation of the root tip. Cells in the growing root divide in a region of tissue known as the **meristem** at the interior of the root (see Figure 2-14). Growth occurs as soon as the seedling germinates. The root that appears from the seedling or embryonic plant is known as the primary root. However, if you have ever noticed a plant or seedling after a few weeks, there is more than one vertical root; these other roots are known as secondary roots, and branch from the original primary root. But roots do not all grow straight in a vertical direction; many parts of the root extend laterally from the primary or secondary roots. These roots are known as lateral roots and originate from the same tissues of the primary root.

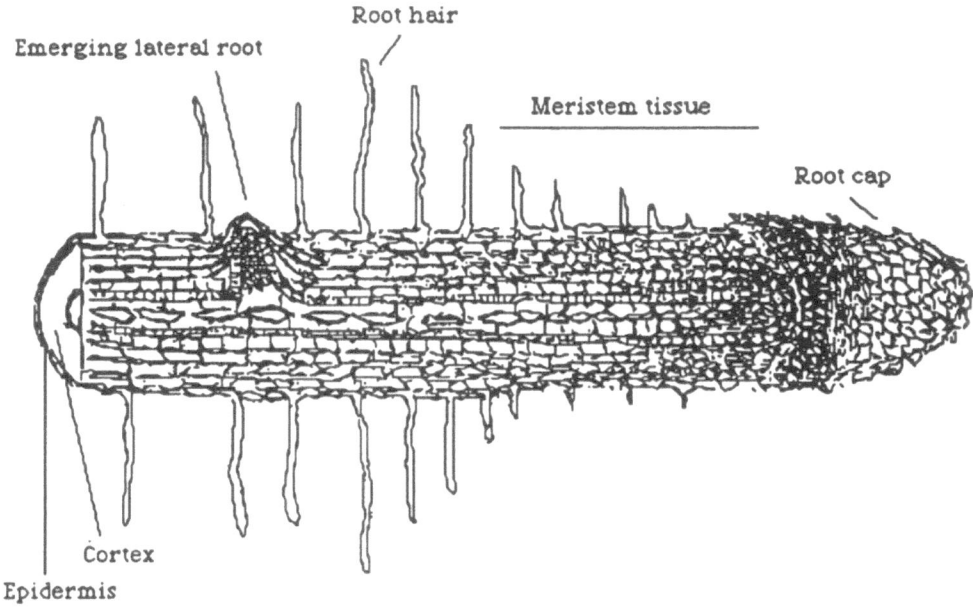

Figure 2-14. Anatomy of a primary root

Many factors affect the growth of plant roots. Of particular interest is the resistance due to the physical barrier of the soil. like the compactness of the soil, can alter root growth. For example, in compact soils, the spaces between soil particles, are reduced in both number and size. The space size is usually referred to as **pore size**. The result is that roots thicken and their rate of growth (vertical elongation) decreases. Their branching patterns become modified and often there appears to be an increase in lateral root growth. As a side note, the lateral roots are often smaller in diameter than the primary root, and are better able to grow through tighter pores. There is much controversy about the full range of mechanisms of soil resistance.

Many other environmental variables affect root growth. In poorly aerated solutions, roots tend to grow straighter, shorter and generate more lateral roots. In corn, light also alters root growth. Blue light inhibits cell elongation and multiplication, while red light inhibits only cell elongation in these species.

Activity 2.16 Growing Roots

2.10.3 Experimental Procedure

In this experiment you will compare the branching patterns which arise when a plant is grown hydroponically with and without fertilizer added to the water. During the experiment you will take photographs of the root system for measurement of the fractal dimension. Finally, once a week for 1-2 months, you will scan or image-grab the root patterns for a fractal dimensional analysis of your root system.

When you record your observations at each date, you should include in your table:

- Length and number of vertical root branches (this includes both primary and secondary root growth).
- Length and number of lateral root branches.
- Number and length of root hairs on vertical and lateral root growth. (These are small and difficult to observe, so use a magnifying glass to observe the hairs)

Refer to the Appendix on Root Growth to build your rhizotrons and germinate your seeds. Any type of seedlings can be used; we have recommended some seeds that have produced interesting root growth. For this experiment it is sufficient to use the stagnant water rhizotrons.

1. Create two groups of rhizotrons and seedlings. Be sure that both groups all consist of the same species of seedling.
   a. Group A - using fertilizer and tap water fill the rhizotron(s).
   b. Group B - same as the control group but no fertilizer.

Fertilizer with the mixture 20-20-20 can be found in any local gardening store. Mix 1 teaspoon of dry fertilizer in 1 gallon of water and apply to rhizotron.

2. Measure and record the root system before you begin the experiment. Indicate the age of the seedling you are using and measure the primary root, secondary root, and root hairs as described above.

3. Carefully place seedling in the rhizotron. (The seedlings are delicate, so be careful not to disturb the roots too much).

4. Try to coordinate some time at least once a week to work with your lab partner making observations and recordings.

5. If you have a video camera, grab an image of your root system through the rhizotron cell. If you have a scanner, take a picture of the root system, and scan the photograph. This should be done periodically, roughly once a week. Follow the instructions in Appendices C or D. Be sure to record your dimensions in your lab notebook and save all your images.

Keep a record of the color and size of the leaves and stem of your seedling over this period of time. You may also consider checking the pH of your solution periodically. Would you expect this to be different in your groups (A & B) and would you expect this to change over time?.

1.10.4 Data Analysis

1. After scanning your images, use the Fractal Dimension program to measure the dimension of your root pattern

2. From your scanned image, plot the length of the longest root of the pattern vs. time and find the root growth velocity. Explain the possible reasons for the shape of your curve. Compare any similarities or differences you observe between groups A & B.

3. Take your notebook observations and investigate the growth velocity from both groups further. Plot the length of the vertical roots vs. time and the number of vertical roots vs. time on separate graphs. Do the same plot for the length and number of lateral roots vs. time.

After a couple of weeks, you may have several root branchings to count. It would be easier to plot this information if you averaged the lengths of the root branchings and averaged the total numbers of root branchings separately for the vertical and horizontal growths. Use these averages to plot on your graph.

4. Now find the growth velocity of lateral and horizontal (primary and secondary) root growth. Explain the similarities and differences you observe or any other observations you had anticipated.

5. How do these values compare with the growth velocity of the root patterns (radius vs. time) you observed?

6. What can you say about root patterns and nutrient solutions? Is there any correlation between the overall patterns you observed
from scanning your images and the manual measurements you made on the different types of root growth in the two different solutions?

7. A number of researchers have written computer programs which simulate root growth under conditions similar to those you have used. How would you use random walkers to create a model of root growth?

See Section 2.8 for further research questions.

2.11 Research!

Now you have the tools needed to carry out your own investigations of electrodeposition experiments and Hele-Shaw experiments. We suggest that you work with a small group and go through the following steps.

Activity 2.17 Research on Electrochemical Deposition

1. Think of a variation of the electrochemical deposition experiment that you would like to carry out, perhaps to investigate some question that occurred to you during the original experiment. These might include

a) Variations of the solution concentration.

b) Vary the salt in solution. Try using $CuCl$ or $ZnCl$ or $AgNO_3$ or tin salts.

c) Try varying the position of the cathode relative to the anode.

d) Try using more than one cathode. How would this compare with the competing bacterial colony growth experiment?

e) Make the spacing between the cell plates uneven. This can be done by inserting a stack of cover slips, or equivalent, in one corner of the cell.

f) If you have a current source, run at constant current rather than constant voltage. See if this results in a change in aggregate shape or rate of growth.

And so forth.

2. Speculate about the results you expect to observe with your proposed experiment. Draw the expected pattern.

3. Submit a written proposal to your teacher, describing the experiment and the expected result. Your teacher will check it for its promise, the availability of components, and safety.

4. After approval, carry out the experiment and compare results with your predictions. Does your outcome suggest further investigations? Where will they lead?

5. Hold a mini research conference in which each group reports on its experiment and interprets the results, allowing time for reactions and comments from the class.

Activity 2.18 Research on Viscous Fingering

You are now a full-fledged research scientist. Propose, get approval for, carry out, and report on an experiment of interest to you concerning viscous fingering. With your Hele-Shaw cell you can observe the viscous fingering properties of a wide range of fluids.

1) Try injecting air into as 50-50 mixture of air and glycerol.

2) Try oil into glycerol.

3) Try sugar water into water, and vice versa. Do you observe anything in this case?

4) Try the Greased Lightning experiment. Take a two plates of plastic and put a spot of lithium grease (available at hardware and automobile supply stores for lubrication purposes). The spot should be about 1/2 cm in diameter. Squeeze the plates just

above the spot as hard as you can. The grease will spread thinly between the plates. Can you predict the shape of the grease as it spreads? Will it finger? Now release the pressure on the grease slowly, and then gradually pull the plates apart. Can you explain the remaining pattern?

Repeat this experiment with a drop of model paint. Or with chocolate frosting.

5) Try putting a grooved plate between the plates of the Hele-Shaw cell and then performing the viscous fingering experiment. Can you predict what will happen?

Activity 2.19 Research on Bacterial Colonies

Vary the agar concentration of the gel. Try 15 g/l (a harder gel), and 7 g/l (a softer gel). What effect would you predict altering the gel will have on the colony pattern? Is your reasoning dominated by questions of nutrient diffusion or bacterial mobility?

Make a diagram of colony shapes in which you plot colony shape versus gel and nutrient concentrations.

Activity 2.20 Research on Termite Colonies

1. Make a mixture of sand with sawdust (say 10% sawdust) and use this to fill the cell. Do the termites form a different pattern?

2. Put a piece of damp paper somewhere in the cell. See what happens if the termites locate this food.

3. Put a thin, damp piece of filter or tissue paper all along the bottom of the cell, and cover it with sand. Does this alter the foraging pattern?

Activity 2.21 Research on Roots

1. Vary the concentration of the nutrient solution by trying serial dilutions, Take your stock solution and dilute it 1/2, 1/4, 1/8, etc..... Based on the research you have done, what effect would you predict altering the nutrient concentration would have on the root patterns? Explain your reasoning.

2. Would you expect aerating the rhizotron to alter the pattern you observed in the stagnant rhizotron? Does this have anything to do with nutrient diffusion? If you have the materials, follow the rhizotron set-up discussed in the appendix, and observe root growth under aerated conditions.

3. Could you speculate how altering different wavelengths of light might affect root growth? How could this cause changes in the plant's response to nutrients?

4. Would you expect any differences in root growth if you filled a rhizotron with sand or soil or vermiculite?

Chapter 2, Appendix A: Dimensions and Logarithms

Why does the LogLog graph work in measuring the dimension of an object? Think of a pattern of fixed area and fixed overall width L. We are going to cover this pattern with boxes of width d and count the number N of the boxes needed to do this covering. For a solid area, we have a dimension of 2 and the general formula.

$$(\text{Area}) = (\text{Constant})L^2$$

The constant depends on the shape. As examples, the following figure shows shapes with three different values of the constant.

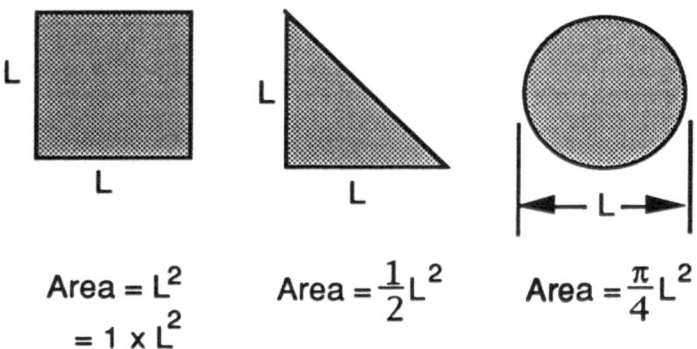

$$\text{Area} = L^2$$
$$= 1 \times L^2$$
$$\text{Area} = \frac{1}{2}L^2$$
$$\text{Area} = \frac{\pi}{4}L^2$$

Figure   Three values of the constant in the area formula.

Now we cover any of these shapes with little boxes of width d and area d^2. How many boxes N does it take. The following formula is approximately correct:

$$(\textit{Area}) = Nd^2 = (\text{Constant})L^2 = (\text{Constant})\left(\frac{L}{d}d\right)^2 = (\text{Constant})\left(\frac{L}{d}\right)^2 d^2$$

or $$Nd^2 = (\text{Constant})\left(\frac{L}{d}\right)^2 d^2$$

Cancel the factor d^2 to obtain:

$$N = (\text{constant})\left(\frac{L}{d}\right)^2 = d^{-2} \times \left[(\text{constant}) \times L^2\right]$$

For a fractal, the dimension is not necessarily 2. Call the dimension D. Then the corresponding equation becomes:

$$N = d^{-D} \times \left[(\text{constant}) \times L^D \right]$$

Now take the log of both sides:

$$\log N = \log\left(d^{-D}\right) + \log\left[(\text{constant}) \times L^D \right]$$

Here L is fixed; we are not changing the area or the overall width L of the figure as we use boxes of different width d to cover it. Therefore everything in the square bracket is a constant, and the log of the quantity in the square bracket is also a constant—call it *const*. Using the property of logs, we have:

$$\log N = -D \log d + \text{const}$$

Think of the variables not as N and d, but rather as log N and log d. Then this equation is that of a straight line with slope, –D. Hence our Log-Log box-covering plots will yield a straight line whose slope is the negative of the dimension D.

Chapter 2, Appendix B: Construction of the ECD and Hele-Shaw Cells

Materials list for ECD

Two 6" x 6" x 1/4" clear acrylic plastic plates. (Dimensions are approximate, due to width of the saw cut.)
Four rubber feet
1 mm Copper Wire
0.5 mm Copper Wire
0.040" drill bit (bit should be equivalent to 1mm or slightly larger)
Four Hose Clamps

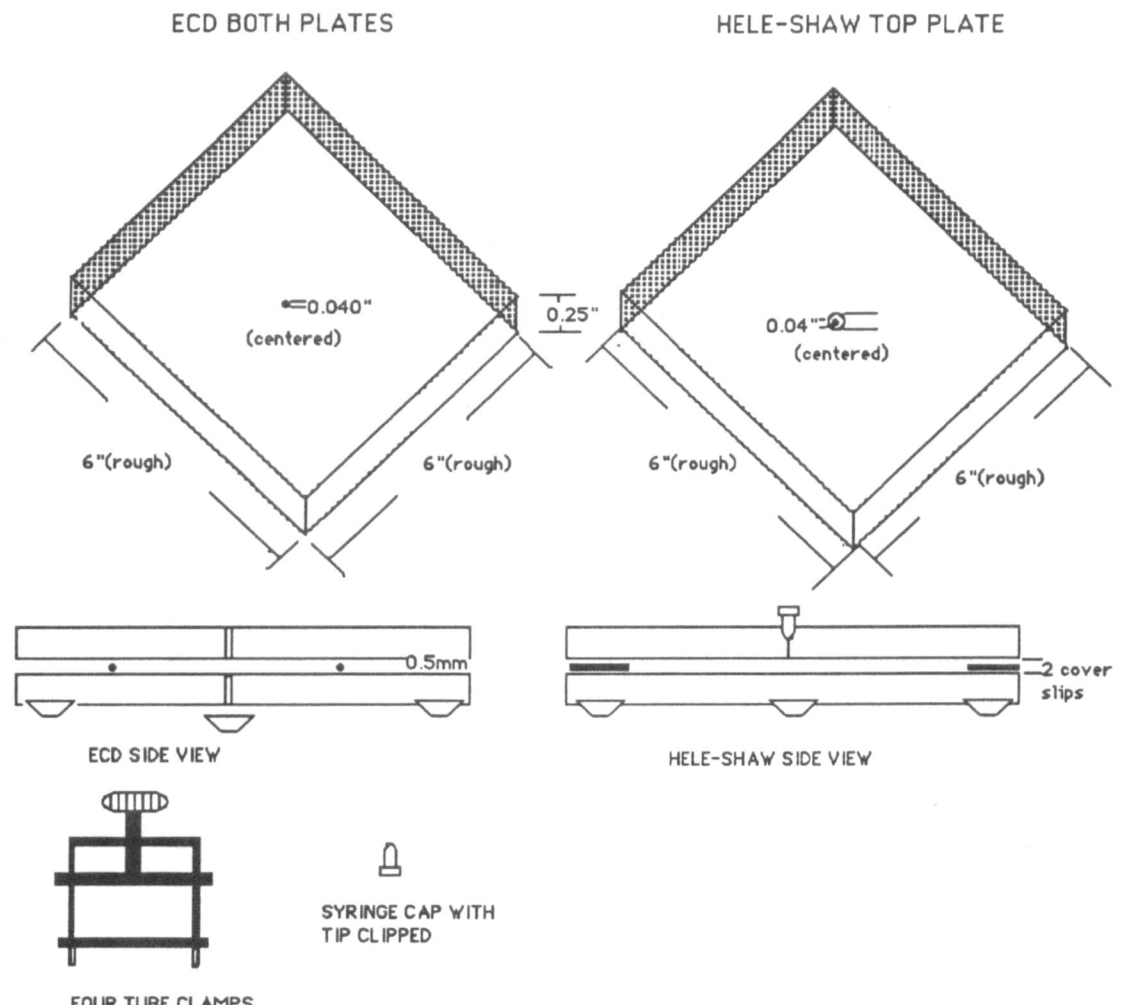

Figure. ECD and Hele-Shaw cell assemblies

Construction of ECD Cells:

Refer to the figure for ECD and Hele-Shaw cells.

<u>Be Safe: Always use safety glasses when working with machinery</u>

1. Take two (2) acrylic plastic pieces and remove the protective paper from all but one face of one square.

2. With a marker draw diagonals on the covered face. The intersection of these diagonals is the center of the square.

3. Securely tape the two pieces of plastic together (covered side on top) with double sided tape (e.g. carpet tape).

4. Using the drill press set on slowest speed, drill a 1 mm (0.04") hole through the center of your plastic squares. Use numbered drill bit 57. NOTE: (a) Place the clamped plastic on a level piece of wood, e.g., plywood. (b) It is best not to attempt to drill all the way through at once. The drill bit melts the plastic as it passes through the plastic and can easily get caught. Drill in, and bring the drill back up, several times to complete your hole.

5. Test the hole to see if a 1 mm wire will pass through the hole. It should be a snug fit so that you don't have to force it in, but neither can it freely fall out. If the hole is too small, drill it slightly larger. NOTE: It is always best to drill small at first and then enlarge. It is easier to remove plastic than to replace it. Due to the imperfections in the thickness of the copper wire, it may be necessary to drill a hole slightly larger than 0.04".

6. Remove the last piece of protective paper.

7. On one face of one piece, affix one rubber foot at the center of each edge. This is now your *Bottom Plate* . The other plate is your *Top Plate.*

Materials for Hele-Shaw Cell

Two 6" x 6" x 1/4" clear acrylic plastic plates. (Dimensions are approximate, due to width of the saw cut.)
Syringe cap from a 20 cc Syringe
0.040" drill bit (bit should be equivalent to 1mm or slightly larger)
0.201" drill bit (bit size may vary, see instructions below)
5 minute Epoxy
Four Rubber feet
Eight Plastic microscope slide cover slips
Two 20 cc plastic syringes
One 1 cc plastic syringe
4 Hose Clamps

Construction of Hele-Shaw Cells

Refer to the figure above for Hele-Shaw cells.

Be Safe: Always use safety glasses when working with machinery

1. Take one (1) plastic plate and remove protective covering from one face.

2. With a marker draw diagonals on the covered face. The intersection of these diagonals is the center of the square.

3. Using the drill press set on slowest speed, drill a 1 mm (0.04") hole through the center of your plastic square. NOTE: (a) Place the clamped plastic on a level piece of wood, e.g., plywood. (b) It is best not to attempt to drill all the way through at once. The drill bit melts the plastic as it passes through the plastic and can easily get caught. Drill in, and bring the drill back up, several times to complete your hole.

4.   Change drill bit to 0.201", and drill a 0.201" inset approximately halfway through the plastic centered on your first hole. DO NOT go all the way through.

5. Take the syringe cap and clip off the tip. Place the narrow end of cap in your inset hole (see the figure). It should fit snugly. If it does not fit, drill the hole slightly larger. Continue this process until the clipped cap fits snugly.

6. After fitting the cap, use 5 minute epoxy and secure the cap onto the face of the plastic piece. **Be Careful** not to seal the 1 mm hole. Allow the epoxy to set overnight to ensure secure bond. NOTE: The best results occur if you mix epoxy on a disposable surface and use something like a pipette or toothpick to apply around the edge of the cap.

7.   Take the other piece of plastic and remove the protective covering from both sides. On one side, affix one rubber foot at the center of each edge. This is now your *Bottom Plate*. The other plate is your *Top plate*.

Chapter 2, Appendix C: Using Video Frame Grabbing and Measuring the Fractal Dimension

1. Go to the Apple Icon on the Finder Menu Bar and select Monitors. In the selection box, highlight Millions (if available) or thousands by clicking the cursor on it.

2. Go to Apple Icon on Finder Menu Bar. Select *Screenplay.*, (copyright Supermac). If screenplay is not available, use the corresponding video capture program on your Mac.

3. Make the Screenplay window larger using the drag point at the lower right corner. There is a limit to the size *Screenplay* allows; make the window as large as possible.

4. Along the bottom of the *Screenplay* window are three buttons. The Start Recording button is labeled with a circle. Initially this circle is red. The Stop Recording button is labeled with a rectangle. This button is initially black but turns blue during Recording.

5. Video clips saved to disk require a large amount of memory. The video camera records at a rate of 30 frames/second. A Mac FX can save to disk a *maximum* of 30 per second. However, it may not meet this maximum. Moreover, if you are saving a long piece, this may not be necessary. To record the video of growth of an electrodeposit (for example) 1 frame/second is probably adequate. For a viscous fingering experiment, by contrast, 20 — 30 frames/second may be appropriate. Depending on your application, set the number of frames/second by going to the Spigot selection on the menu bar, select Preferences, and at the box labeled "Limit frames per second to" set the frame rate. Click OK.

6. Record the image using *Screenplay*. Click the Record Button (with the red circle) to start recording, and then click the Stop Button (the blue square) to stop recording. This may be very brief if all you want is a still image of a final electrodeposit growth. Long clips (greater than 10 seconds at 30/sec) will start pushing the FX's memory.

7. After you stop recording, a new window will appear labeled Untitled.

To save a still with Screenplay, it is necessary to first save the clip as a GrayScale image. Saving the clip as gray scale to disk requires a significant piece of memory. For this reason, first edit the clip to minimize its length.

8. Using the thumbscrolls on the bar directly below the image you can select a segment from the video clip. The middle thumbscroll advances the video frame-by-frame. The outer left thumbscroll can be moved to the right to set the beginning of the segment. The outer right thumbscroll can be moved to the left to set the end of the segment. Select a short segment which contains the frames which you want to analyze. (N.B.: Do not put the left and right thumbscrolls against each other. There must be some space between them. If the segment of video is too short, then in the next step the Save As option will not be available.)

9. Under the File menu option select Save As and specify a name for your segment in the dialog box. Click on the Compression option. In the new dialog window, the upper pull down menu (titled Compression Method) should be set to Graphics. The Colors pull down menu should be set to GrayScale. The quality scroll bar should be set to High. Click O.K. And then click on Save.

At this point the video segment on your screen will lose color and be gray with appropriate grayscaling.

10. Run the gray scale video to the frame you wish analyze. With the cursor click on that frame. The cursor will immediately convert to a Hand icon. With the Hand move the frame image off of the video window. The result will be a new window which is a Still image.

11. From the File menu option, select Save. This will save the still.

12. Quit Screenplay. This is done because the applications to be run below require substantial memory, as does *Screenplay*.

****

Image Manipulation with Image 1.44

1. Open Image 1.44 under the Apple icon. If you haven't already done so, Image 1.44 will launch a dialog box reminding you to revert to 256 colors. This you do by selecting Monitors under the Apple icon, and selecting 256 colors.

2. Under File menu option select Open, and find your saved image on the hard drive.

3. Under the Options menu option, select Threshold. Release the mouse. Your image will appear with a different contrast. Most likely it will not be contrasted the way you would like. Don't panic.

4. Under the tools palette on the left, you will find the threshold icon highlighted. Move your cursor to the vertical thresholding LUT to the left of the tools palette. Move the demarcation between white and black with your cursor. As you do this, you will observe that the thresholding of your image alters. Find the best thresholding.

At this point you can clean up your image using the **eraser** in the tools palette. The eraser is very similar to the one in MacPaint.

5. Under the File menu option, click on **Save As** and title your document. (You will have to change the name of your file.) Select *MacPaint* from the push buttons for your save format. Click on Save.

7. Close Image 1.44.

****

Measuring the Fractal Dimension

You now possess a *MacPaint* file that can be analyzed using *Fractal Dimension,*.

1. If you are using the FX, check that your monitor is set for 256 colors. Go to the Control Panel and select Monitors. Under Monitors select 256.

2. Open *Fractal Dimension*.

3. Open saved *MacPaint* file.

4. Proceed with directions given in program.

5. Be aware that depending on the morphology of your image, you may want to use the **Circle Method** found in the **Display/Method** pull down menu.

Chapter 2, Appendix D: Using the Scanner for Imaging and Measuring the Fractal Dimension

PRELIMINARIES:

a.) After running the electrochemical deposition experiment, carefully wipe the exterior of the cell with a paper towel to dry it. Do this with care so as not to damage the aggregate.

b.) Remove the cathode wire (i.e., the wire inserted in the center of the cell).

USING THE SCANNER:

1.) Place the electrochemical deposition cell on the platen with white paper over the deposit for contrast. Also cover the remainder of the platen with white paper.

2.) Check that the memory for the application AppleScan is set at 2000 K.
Open AppleScan.

3.) Initial values of Scan Control Panel can be used for: the resolution pull down menu (300 dots per inch); the line art button; and for the Threshold in the settings pull down menu. Click on the Reduce/Enlarge button. At the top of the pop-up dialog box there appear a sequence of page sizes. 100% is to the far right. With the cursor a little beneath a page near the center of the sequence, click the mouse. A new size % value will appear. (The response to the cursor is sensitive to distance beneath the page where the mouse is clicked. Experiment until you succeed.) Select the page size for 55% reduction. Click on the OK button.

4.) Click on Scan button.

5.) **Close Scan Control Panel window. Bring to the foreground the window titled "Untitled - 1". From View menu select Dot for Dot. With the thumb scroll bars, center the aggregate in the window.** This can also be accomplished using the hand icon from the tools palette. (Warning: with the hand icon the image may continue drifting after dragging is seemingly completed. Click on the mouse to stop the drift.) The aggregate should not be larger than 9 cm in diameter on the screen or it will be too large for the fractal application to analyze. The image can be reduced by selecting ACTUAL SIZE from the View menu.

6.) Press simultaneously the keys SHIFT-CONTROL-C on the keyboard, and use the mouse to drag a box around the aggregate. This puts the image in the clipboard.

7.) From FILE menu select New file.

8.) From EDIT menu select Paste.

9.) To save as a MacPaint file:
     a. From File menu select SAVE AS. Select MacPaint from dialog window.
         b. Title image if desired. Select SAVE.
         c. From the File menu select QUIT. Do not save changes to UNTITLED.

10.) Open Fractal Dimension. Click once to erase credits.

11.) From the File menu select Open. From the dialog box open the AppleScan folder, or the appropriate folder where your MacPaint image is stored. Select your image.

12.) Read the instructions in the message box. Click the mouse to close the message box. Following the instructions provided, select a box to be dragged around the image. Drag the mouse so that the box frames the image and click at the lower right hand corner of the image to lock the frame. Select the appropriate button from the new message box in the frame.

13.) Determine the fractal dimension by the box counting method.
     a. Select the number of pixels per box from options buttons.

b. Wait 15 seconds for the initial scan.

c. Select additional buttons for additional data points.

   d. To drop a data point from the line of best fit, click the cursor on it.

14.) Determine the fractal dimension by the circle method.

   a. Under options menu select circle method from the right hand column. Click OK.

   b. Choose manual center and place the center at the point that

      the growth originated from.

   c. Drop the data points that are far from the line of best fit.

Chapter 2, Appendix E. Preparing Carrageenan Solutions

This appendix describes how to prepare carrageenan solutions of differing concentrations and pH. Carrageenan is available from chemical supply companies (e.g., Sigma).

WARNING: Wear safety glasses and gloves throughout this preparation.

1) Fixing the concentration of carrageenan solution. Carrageenan comes as a powder. It is slow to dissolve in water. Be prepared to let the solution stand for at least 24 hours to permit the powder to dissolve. The resulting solution is also an excellent medium for growth of bacteria and mold. For this reason it must be used within a few days of preparation.

To make a concentration of 5 mg/ml, add 0.5g of carrageenan to 100ml of distilled water. To make a concentration of 10mg/ml, add 1g of carrageenan to 100ml of distilled water. Cover the solution (if available, parafilm is excellent for this purpose) and leave it overnight for the powder to dissolve. After the carrageenan has dissolved, mix the solution for a few seconds to making it homogeneous.

2) Fixing the pH of carrageenan solution. Initially the carrageenan solutions (both concentrations above) will have a pH of between 8 and 9. Prepare a dilute solution of HCl acid (e.g., 2 ml of HCl (36.5 —38%) in 100ml of distilled water). Add the diluted HCl drop wise to your carrageenan solution, mix the solution, and using a pH meter, measure the pH of your

solution. By repeating this process, prepare carrageenan solutions of pH 3 and 5.

3) <u>Preparing HCl at pH 3 and 5:</u> Add 1 ml of HCl to 100 ml of distilled water then take it's pH. Then change the pH by adding distilled water to arrive at the desired pH. (Distilled water can have a pH of 5. For this reason, starting with a very dilute HCl solution is necessary.) Since the experiment that follows is **not** dependent on chemical purity, tap water (pH 7) can be used to bring the pH up.

Once you have the pH you want, add food coloring to the HCl solution so the pattern formed by the HCl in the Hele-Shaw cell can be observed.

Chapter 2, Appendix F. Using the Vernier pH Meter

CAUTION: Never let the bulb of the pH meter become dry. If it does, the pH meter will be broken.

1) Connect the Vernier Universal Laboratory Interface (ULI) to the modem port (labeled with a telephone icon) at the back of the Macintosh.

2) Connect the pH meter to the ULI in either DIN1 or DIN2 (corresponding to Ports 1 and 2 in the Data Logger).

3) Find the application called Logger V1.31 and start it.

4) In the Data Logger window a graph will appear. Go to Collect and click on Calibrate.
5) Click on Calibrate now.
6) Choose which port you are using.
7) The directions on how to calibrate will appear. Use known buffers to calibrate.
8) For 'stable reading', type in the pH of the buffer you are using then click on OK.
9) The graph will reappear. Click on Start, place the probe in the carrageenan solution and read the pH reading. Remove the probe.
10) To adjust the pH, add the HCl solution (or a more basic solution to increase pH) drop by drop then mix with stirring rod.

11) Return the probe to the solution. Between each reading you should rinse the probe with distilled water to assure that the pH reading is accurate.

Chapter 2, Appendix G: Preparation of Bacterial Growth Experiment

MATERIALS:

This experiment requires the following supplies and equipment:

Equipment: incubator (if available) and autoclave.

Lab supplies: petri dishes, beakers, aluminum foil, autoclave gloves, beakers, inoculating stick, inoculating loop, parafilm.

Chemicals for nutrient agar: Bacto-Agar, Bactopeptone, potassium phosphate dibasic, NaCl, concentrated HCl, and distilled H_2O.

Staining supplies: 0.1% Coomassie Brilliant Blue R stain (Sigma Chemical Co.), methanol, vinegar (or acetic acid), distilled water.

Bacterial sample: B. subtilis or E. coli or E. aerogenes.

Bacteria can be purchased from the Carolina Biological supply company. Specific bacteria can sometimes be obtained from stock centers. B. subtilis is available from the stock center at Ohio State University and E. coli is available from the stock center at Yale University. Both stock centers provide the bacteria for free.

Preparing the Agar Plates

1. Bacteria are grown on nutrient agar. To vary the nutrient levels in this experiment, we vary the bactopeptone levels in the solution. For example, below are listed the recipes for nutrient concentrations of 10 g/l, 5 g/l, 1 g/l, and 0.5 g/l. To prepare volumes smaller than a liter, scale down the quantity of each ingredient by the same ratio as the volumes.

Recipes:

	10 gm/liter	5 gm/liter	1 gm/liter	0.5 gm/liter
distilled H$_2$0:	1 liter	1 liter	1 liter	1 liter
bactopeptone:	10 grams	5 grams	1 gram	0.5 grams
Bacto–agar:	10 grams	10 grams	10 grams	10 grams
K$_2$HPO$_4$:	5 grams	5 grams	5 grams	5 grams
NaCl:	5 grams	5 grams	5 grams	5 grams

2. Place the ingredients in beakers, one beaker for each concentration. Also place in each beaker a glass stirring rod to be sterilized. You will need this rod after autoclaving to stir your solution. Put a cover of aluminum foil on the beakers (shiny side facing the inside of the beaker), and put the nutrient agar in an autoclave. The time necessary to sterilize the solution depends on your particular autoclave.

3. After autoclaving for the specified time, remove the beakers wearing autoclave gloves. Remember that the autoclave operates at 250°F, and a pressure of 15 atmospheres. The solution which comes out is **very hot**. *Be careful!*

After removal from the autoclave, lift the aluminum foil at one corner to create a steam release path. The foil should remain up in that corner (roughly 1/8 the circumference of the beaker). After roughly 15 minutes the solution has cooled sufficiently (a temperature of 90 - 95°C) to be used. Leave the glass rod in the beaker during this time.

BE CAREFUL!! In the following, you are using an extremely dangerous acid. Wear gloves.

4. Adjust the pH of the solution to be 7. Using a dropper, add **one** drop at a time of concentrated HCl and stir with the glass rod already in the beaker. To test the pH, lift the glass rod out of the solution, and allow a drop to fall onto litmus paper. *To avoid contaminating the rod, do not touch the rod to the paper*. It may take about 5 drops of concentrated HCl to reduce the pH of 100 ml from 8 to 7.

5. Pour 20 ml of nutrient agar (or until the agar covers the bottom of the dish) per petri dish. Cover immediately and allow the agar to solidify (fully gel). This takes a minimum of two hours. At this point turn the plates upside down and dry for 48

hours at room temperature. If the plates are not inverted during drying, condensate accumulates on the top plate and drips onto the agar surface. This can result in an irregular agar surface. Drying eliminates excess water in the agar and reduces the condensate, which can affect bacterial colony patterns.

Inoculating and Incubating the Plates

1. Light a Bunsen burner in your bench space. To maintain sterile conditions, inoculation should occur within 20 cm of the flame. Wait 20 seconds before opening the petri dish and inoculating. This gives the flame time to sterilize the local air. (N.B.: The idea here is to achieve sterile conditions. This will be violated, despite the fact that the Bunsen burner is on, if you lift the plate and work with it close to your face!)

2. To sterilize the inoculating loop, hold its tip in the flame until it turns red.

3. Poke the inoculating loop through the agar close to the side of the petri dish to cool it. This prevents the heat from killing the bacteria sample you want to use. The heat will not harm the agar.

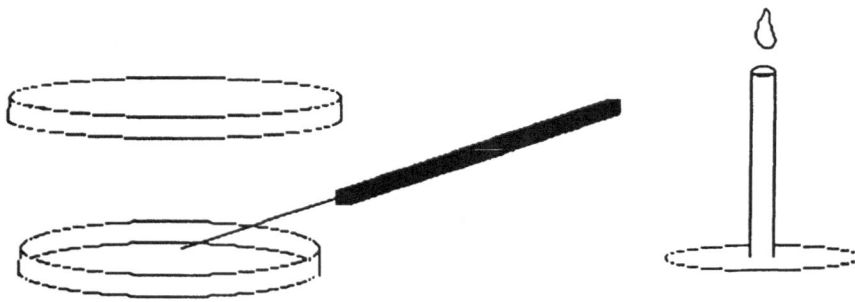

Figure. See instruction 4

4. Touch the inoculating loop to a single colony of bacteria on the sample plate. Turn the petri dish over and use a permanent marker to make a dot on the under side of the petri dish, marking the exact point at which you wish to inoculate. Now touch the agar in the petri dish above the mark with your needle. A little dab will do you; *don't poke the needle into the agar*. You have now inoculated your bacteria sample onto the petri dish.

5. Close the petri dish and seal it with a layer of parafilm around the edge. This keeps the agar from drying out while it is in the incubator.

6. Incubate the sample at 37°C. Make sure to keep an open beaker of water in the incubator. Periodically check that the beaker has water in it -- do not let it run dry. The water will maintain a constant level of humidity (100%) in the incubator. Despite your best efforts at parafilming, there is always some leakage.

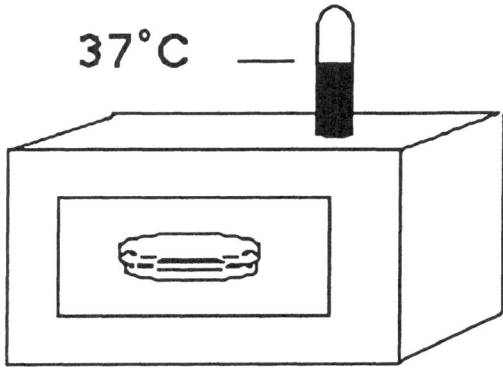

Figure. See instruction 6

Staining

When you want to preserve your bacterial colony, you can stain it. Staining the colony kills the bacteria to prevent further growth, and makes the colony easier to look at both with the naked eye and under a microscope.

1. *Staining Solution:* Prepare a solution of approximately 0.1% stain in a beaker containing a liquid made up of the following proportion of constituents by volume methanol: vinegar: distilled water (50:10:40). (If acetic acid is available, use this in place of the vinegar.) When making up the solution add the water first, the methanol second, and the acid last. If the total volume is 1 liter, add 1 gram of Coomassie Brilliant Blue R stain powder. Use proportionately less for smaller volumetric quantities. Mix the solution until it is a uniform blue color.

2. *Rinse Solution*: Prepare another solution of (50:10:40) of methanol: vinegar: water without the stain. Mix well.

3. Remove the parafilm and lid from the petri dish. Pour the solution containing the stain into the petri dish until the surface is covered. Gently swirl the solution in the petri dish for 60 seconds, no shorter or longer period of time. Pour out the excess solution from the petri dish. **Take care not to pour out the agar which may now separate from the petri dish.**

4. The plate surface must now be rinsed of excess stain. With the solution prepared in Step 2, cover the surface of the plate as you previously did with the stain. Immediately pour off the rinse solution. Repeat this rinse procedure (pouring on and off the rinse solution) four or five times, or until the desired contrast of stained colony to background agar is achieved. Replace the lid and carefully parafilm the plate again to preserve it.

****

REFERENCES: There is substantial interest in the current physics literature in studying bacterial colony forms. See M. Matsushita and H. Fujikawa in Physica A168, 498 (1990); also E. Ben-Jacob et al in *Pattern Formation in Physical Systems and Biology*, ed. P. Meakin, L. Sander, and P. Garcia (Plenum, in press). The growth forms of *P. Mirabilis* has been extensively studied by James Shapiro at the University of Chicago. For example, see Physica D49, 214 (1991).

Chapter 2, Appendix H Constructing the Cell for a Termite Colony

Obtaining a Termite Colony

Finding wild termites is an exercise which will vary by region. In New England we go to a wooded park and search among the remains of dead trees on the ground. Termites like soft, damp wood. An ax, or at least a hammer with a good claw, or a crowbar, are useful tools to pry open, or split open rotting wood. You will encounter many different insects, in particular carpenter ants, but with luck you will find termites. The best time for this is in warm weather. In the fall termites go underground for the winter. (What can be so fine as a day in June hunting termites with your hatchet?) The termites we found locally are the species flavipes reticulitermes.

Collect pieces of wood with termites in plastic bags (trash bags work fine) and bring them back to the laboratory.

Maintaining Your Colony

In the lab, wood with termites can be kept in a large plastic pan covered by a plastic sheet/garbage bag. Periodically the wood should be sprayed with water to maintain an adequate moisture level. The greatest threat to the colony is dehydration: the cuticle of a termite is thin, and dessication can occur quickly. In the wild, termites foraging activity is highest in the rainy season, and at night, when the relative humidity rises.

Materials for Termite Cell Construction:

>   2 sheets of Plexiglass 12"x12"x3/8"
> plastic shim   12"x12"x1/32" (1/2" border width)
> Weldon 3 solvent cement for joining Acrylic
> 4 tubing clamps
> 4 C-clamps
> cotton balls
> sand-small granular typically used for cement mixture
> acetone
> petri dish

Refer to the figure for the following steps.

1. Tear off the paper covering the Plexiglass sheets. Take care not to scratch the plastic as this will interfere with viewing the termites. This will become the inside of your cell, so be careful not to scratch the surface.

2. Using the acrylic cement, attach the plastic shim around the edge of one of the plastic plates . This is your bottom plate.

3. Drill a hole approximately 1/2 inch in diameter in the center of your top piece and the bottom of the petri dish.

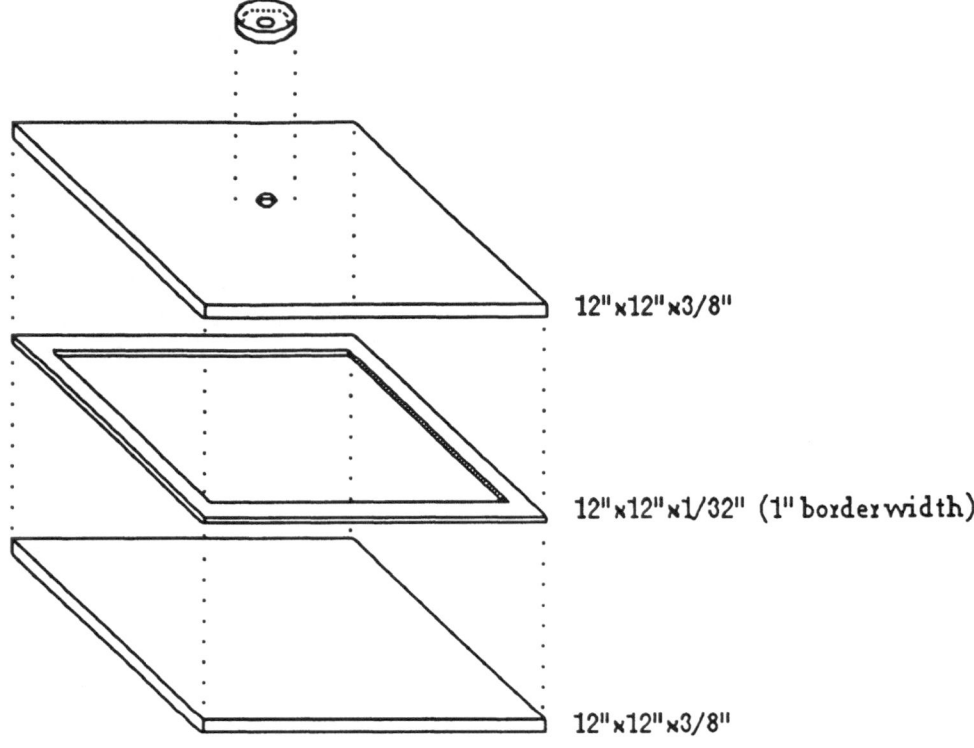

12"×12"×3/8"

12"×12"×1/32" (1" border width)

12"×12"×3/8"

Figure. Termite cell assembly

Cell Set-Up

1. Moisten a cotton ball with acetone and lightly wipe the inner surface of the cell. Allow a few minutes for evaporation. This will remove fingerprints and organic residues.

2. Fill a beaker or other container with sand and slowly the sand sift over the bottom of the cell until the surface is evenly covered. Level the sand to 1/32 in. [HINT: This can be accomplished by using your Plexiglass top as a straight edge to smooth it out. Hold the top sheet perpendicular to the cell and slowly drag it across the surface. You will know when the sand is flat because very little sand will be displaced as the top is dragged along the cell.]

3. Moisten the sand with the water using an atomizer. The sand should be wet but not so wet that pools of water accumulate on the sand. About 1 fl oz. of water is the right amount. (After the cell is closed, by the second day condensate may form on the interior of the plates. That is a good sign.)

4. Secure the top of the cell to the bottom using clamps. First secure with the 4 tubing clamps on the corners, and then with the 4 C-clamps in the middle of each side.

5. Drill a hole in the center of the bottom of the petri dish to match that of the top sheet of Plexiglass. To minimize obstruction of the pattern, use the smallest petri dish available.

Inserting Termites into the Cell

1. Using clear tape, secure the bottom of the petri dish to the top cell such that the holes line up.

2. Break open a potential piece of wood over a container. If the termites are present, they will fall harmlessly in to the container. Repeat this process until you have shaken loose the number of termites you need, roughly 100 to 200 termites.

3. With a pair of termite forceps transfer the termites, one by one, to a petri dish.

4. Pour the termites gently into the petri dish on the cell, close the dish's lid, and seal its edges with tape.

5. To take pictures, place a sheet of black paper on the bottom of the cell for better contrast.

Removing termites from the cell

1. After the completion of the run, unscrew and remove the top plate of the cell. Almost all the termites will remain on the sand.

2. The sand should still be sufficiently moist that if you incline the bottom plate nearly vertically, the termites will roll out of the cell but the sand will remain stuck to the plastic. Do this over the wood/dirt where you maintain your breeding colony of termites. The few termites remaining can be removed with the forceps.

Cautions to the Researcher

1. Always wear gloves when handling the Acetone and acrylic Weld On 3. Both of these are carcinogens.

2. Acetone and joiner produce noxious fumes. Do not inhale.

3. Always wear safety glasses during the set up stages.

Cautions for the Termites

Handle the termites and the cell with care. Avoid excess lighting
on the cell so as not to cook the termites. Even in the absence of
supplied food, the colony should prosper for up to two weeks.
The most important ingredient to colony longevity is moisture.
If the sand appears dry, the cell should be emptied.

Chapter 2, Appendix I. Constructing Rhizotron for Root Systems

Reminder: A **rhizotron** is a clear-walled chamber through
which one can observe roots as they grow.

MATERIALS:
For Germination
    petri dish
    filter paper or white paper towel
For Rhizotron
    2 Plexiglass sheets of equal size about 30cm by 15cm,
    and 1.5cm thick.
    1 Plexiglass sheet about 30cm by 10cm, and 1.5cm
    thick.
    2 Plexiglass spacers of equal size- should be as high
    as the cell, and anywhere from 2mm to 1 cm wide
    acrylic sealant- Weld-on 3
    Plastic Tub- can be purchased, or constructed from
    Plexiglass.
    Fish tank bubbler (about $10.00)
    Seeds (peas, squash, sunflowers, etc.- can be
    purchased from the Burpee catalog.)
    Nutrient solution - (Peter's Professional Fertilizer
    20-20-20, about $10.00)
    Black garbage bags

SEED GERMINATION
The seeds must be germinated before they are placed in the
rhizotrons. Germination allows the seeds to grow a large
enough root system so that they can be supported by the top

edges of the rhizotron frame. This allows the seed to sit on top of the nutrient solution, supported by the frame, with the roots in the solution and the young shoot of the plant extending beyond the top of the frame into the air. Prior to germination, seeds are too small to be supported by the plastic and will sink to the bottom and drown.

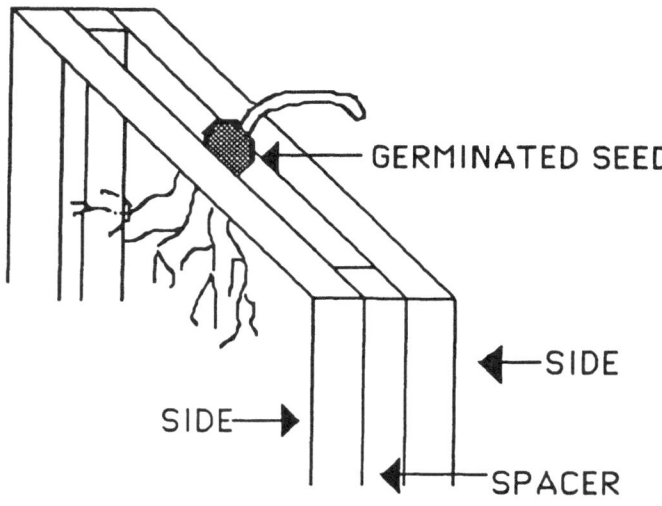

Figure. Seed placement in the rhizotron

1. Mix nutrient solution in the proportion of 1 teaspoon Peter's Professional fertilizer to 1 gallon of water.

2. Soak a few sheets of white paper towel or filter paper in water.

3. Place some of the wet sheets in the bottom of the petri dish, and place the seeds in the dish on top of the paper.

4. Cover the seeds with more wet paper and put the cover of the petri dish on. Store in a warm, dark place. You will want to cover the petri dish in a black garbage bag to insulate it and help keep out light.

5. The seeds and paper must be moistened daily with water.

6. The seeds should be allowed to germinate until they are large enough to sit at the top of the rhizotron frame, supported by the plastic sides (figure). The following is

a list of average germination times for commonly used plants:

Peas- 3 days
Beans- 4 days
Sunflowers- 4 days
Barley- 4 days
Tomatoes- 5 days

PROCEDURE: There are two ways to run the experiment from this point. In the first method the roots receive more oxygen, and therefore grow at a quicker rate. When oxygen is not supplied to the plants, stunted growth and even plant death occurs. One option is to run both versions of the experiment side by side so that variations in growth can be observed.

Method 1: Root Growth with Oxygen

WEAR SAFETY GLASSES FOR ALL CUTTING OF PLEXIGLASS!
During cutting, dust and sharp plastic shards are produced that are extremely dangerous to the eyes!

CONSTRUCTION OF THE RHIZOTRON FRAME: (see figure)
1. Cut the Plexiglass to the dimensions specified in the materials section. All sides and bases should be the same size regardless of the type of seed used. The spacer width depends on the type of seed you have decided to use. The following is a list of spacer widths for commonly used seeds:

Peas- 3 mm
Beans- 9 mm
Sunflowers- 1 cm
Barley- 3 mm
Tomatoes- 3 mm

2. Drill a minimum of 4 holes in the spacers of the rhizotron so that oxygenated water from the bubbler can flow through the rhizotron, and get to the roots of the plants. The holes should be large enough to allow water to flow through the rhizotron adequately. The larger and more numerous the holes are, the better the experimental results.

3a Glue both spacers to one side, making sure that the bottom edges are flush.

3b. Glue the other side to the spacers again being sure that the bottom edges are flush.

4. Glue the base of the cell on to one end of the previously assembled sides and spacers. The base of the cell has to be glued on carefully so that the rhizotron stands upright.

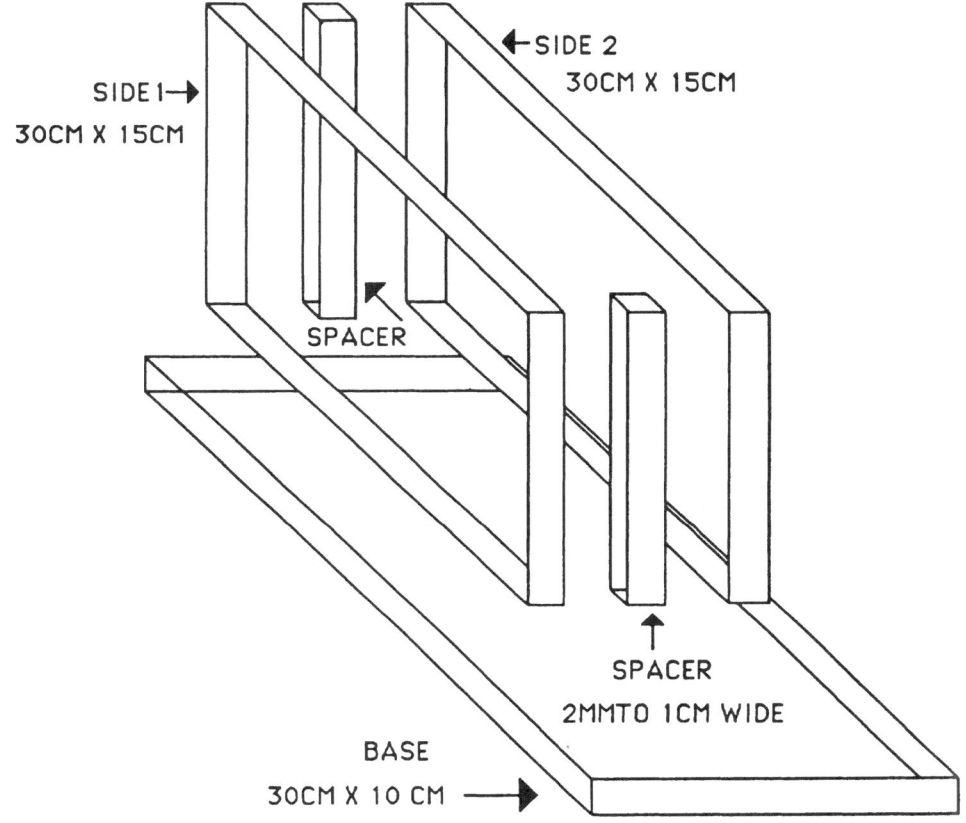

Figure. Exploded Assembly Diagram of the Rhizotron

PLASTIC TUB AND NUTRIENT SOLUTION:
The bubbler in the plastic tub circulates water around the tub and through the rhizotrons. In this way the roots of the plants receive much more oxygen then they would by simply sitting in a rhizotron that was not connected to a tub and bubbler unit.

1. Fill the plastic tub with nutrient solution so that it covers the roots adequately. (see figure)

2. Place the bubbler in the tub. The bubbler should hang over the edge of the tub so that its motor is on the outside and the oxygenation unit is on the inside of the tub, slightly submerged.

3. Place the rhizotrons in the tub.

Figure. Integration of Rhizotron and Bubbler Unit

PLANT INTRODUCTION:

1. When the seeds have germinated, sort through them and pick out the largest seeds, or the seed that will best be able to sit on the top of the frame.

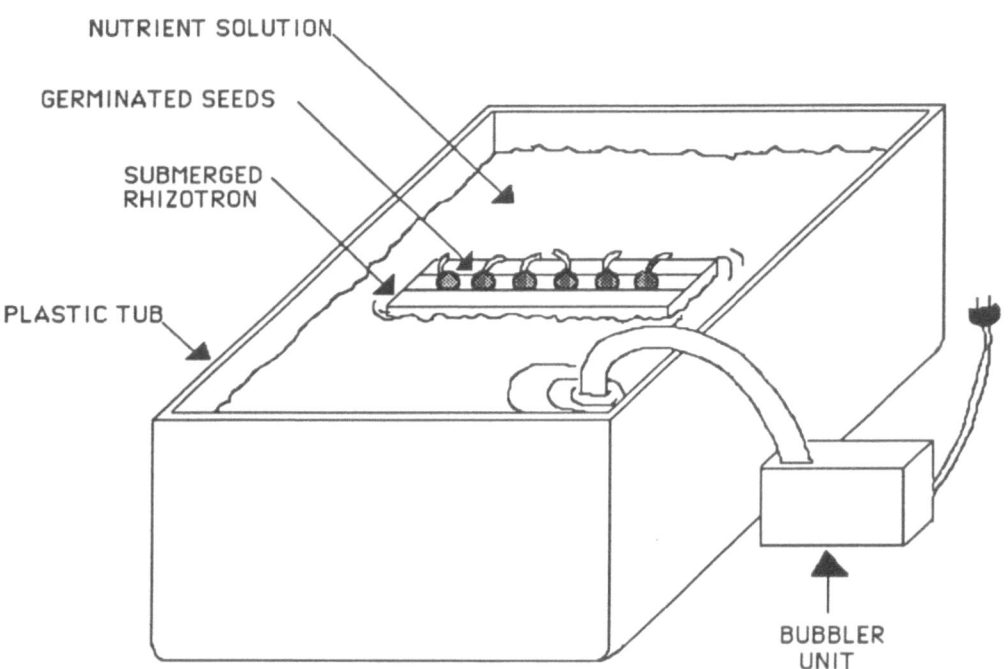

NUTRIENT SOLUTION

GERMINATED SEEDS

SUBMERGED
RHIZOTRON

PLASTIC TUB

BUBBLER
UNIT

2. Fill the rhizotron with the nutrient solution to the top of the frame.

3. Place the selected germinated seeds carefully on the surface of the frame.

4. Carefully tuck the roots of the seed in between the sides of the frame. NOTE: the young shoot of the plant should be sticking out above the top of the frame into the air.

Method 2: Root Growth in Stagnant Solution

CONSTRUCTION OF THE RHIZOTRON FRAME:

Identical to Method I except omit step 2 of Method 1. After assembly, fill the frame with water and let sit for a while to make sure the frame doesn't leak.

PREPARATION OF THE RHIZOTRON:
1. Fill the rhizotron with nutrient solution so that the solution is level with the top edges of the rhizotron.

2. Wrap a strip of black plastic garbage bag around the rhizotron. Make sure that the top edges of the plastic do not extend past the top edge of the rhizotron.

****
*

REFERENCES

Root anatomy/physiology
Feldman, L.J. (1984). Regulation of root development. *Annual Review of Plant Physiology* 35: 223-42.

Campbell, N. A., *Biology*. Menlo Park, The Benjamin Cummings Publishing Co., Inc., 1987.

Computer modeling of root structure
 A.H. Fitter *et al.* (1991). *New Phytologist* 118: 375.

Contents

Chapter 3. Growth Patterns in Nature: Percolation

Chapter 3

Growth Patterns in Nature: Percolation

3.1 Growing a Forest

3.1.1 Introduction

Think of a forest. The forest grows when an existing tree releases its seeds and a new tree grows next to it. The new tree grows and seeds a tree next to it. And so on. The forest burns when fire spreads from a burning tree to its neighbor. The spread, either of new trees or fire, to adjacent sites is an example of **percolation.** Will the forest spread, continuing to grow? Will the fire spread, destroying the entire forest? Important questions for the forester and the environmentalist! The answers? They depend on the chance that a new tree grows next to an existing tree or the chance that flames spread from one tree to the next tree. When we try to predict (or at least study) what might happen, we must consider the probability or chance that the event will occur. An event that has a zero probability of happening will *never* occur. Another event that has probability unity (or a probability of one) will *always* occur. Most probabilities have a value that falls between zero and one. For the forest, what might affect the probability that seeds spread from one tree will take root and grow another tree?

Low probability that a new tree grows next to an existing one? Then the forest may stop growing. Low probability that flames spread from a burning tree to the next one? Then the fire may not spread at all. High values of these probabilities? Then the spread is complete and rapid: the forest grows densely or burns down completely. These are the extremes.

Between extremes, between low values of probability and high values, a strange thing happens. For probabilities below a certain "critical" value, the forest usually does not grow (or burn down); above this value the forest almost always grows (or is consumed). The change takes place at or near the *critical probability*. At the critical probability, we say we have reached the *percolation threshold*. The behavior of the system changes dramatically when the probability passes the percolation threshold.

Forest growth is only one example of critical phenomena. Here are others:

- Red-hot iron is not attracted by a magnet, even though each iron atom is itself a little magnet. Thermal agitation—the erratic jiggling of atoms due to heat energy—keeps the atomic iron magnets pointing every which way. Now lower the temperature while holding the iron in a magnetic field. Thermal agitation drops with temperature. Below a critical temperature the iron atoms line up, each held in direction by the influence of its neighbors. Remove the iron from the magnetic field; the block of iron has become a permanent magnet.

- A gel (example: "Jello") is a "wiggly" solid composed mostly of water contained in invisible "bags" made of hooked-together gelatin molecules. For low gelatin content the result is a liquid; there is not enough gelatin to form bags. Gradually increase the fraction of gelatin; at a **critical concentration** of gelatin the liquid stiffens into a gel.

- Stars in galaxies burn up their fuel by nuclear reactions, then collapse, and explode, scattering their contents into surrounding space. From this scattered dust new stars form due to gravitational attraction, starting the cycle again. How dense does the dust have to be for the creation of new stars? Below a certain **critical density** the galaxy dies away because new stars are not formed. Above this critical density new stars replace old stars and the galaxy continues to exist.

Activity 3.1 Growing a Forest by Hand

The forest has burned down entirely. A single tree in the center reseeds the forest, dropping a seed next to itself. There is some probability that the seed will grow, and some probability there will be a rock where the seed lands, so that seed will not grow.

The tree in the middle drops more than one seed. Each seed either hits a rock or grows. Time passes. Each growing seed becomes a tree. Now each new tree drops seeds on new locations near it, with the same probability of success for each seed. And so it goes; each tree grows and scatters seeds. The forest may grow to fill in all available space. Or maybe the growing cluster of trees will be stopped by rocks all around.

The Forest activity and computer program are designed to encourage interactive exploration of percolation and its results. In the initial activity, you grow a forest by using coin flipping to understand the basic process.

Work in pairs, each pair will need a checkerboard, a set of red and black checkers, and a handful of pennies. Before beginning the procedure below, make a prediction:

Predict!

Now stop for a minute and make some predictions. What do you expect to happen? What will your pattern of trees look like when it is finished? Will your pattern look exactly like everyone else's pattern of trees? As you go on, do you think your forest will reach the edge? What would keep it from reaching the edge?

Now carry out the following steps:

1. Place a red checker on a square near the center of the checkerboard. This stands for the tree whose seeds start the forest growth.

2. Shake up the pennies between your cupped hands and, without looking at them, put one penny in each of the four squares right, left, above, or below the seed tree (not on the diagonals). For example:

3. Look at the four pennies. A head means a new tree; a tail means a rock. Replace each head with a red checker, each tail with a black checker.

4. Shake the pennies up again and place one in every empty square next to a red checker (not on the diagonals).

5. Again replace heads with red checkers, tails with black checkers.

6. Repeat steps 4 and 5 until there are no more open spots next to trees or until one tree in your forest reaches the edge of the checkerboard.

On the board in front of the class, draw a table with two columns. Label the first column "Reach edge" and the second column "Not reach edge." As each pair of students completes growing the forest, they report on whether or not their pattern reaches the edge. Enter their result in the table. Then each pair starts a new pattern from the beginning. When each group has completed three patterns, stop the activity and look together at your table on the board. Have a discussion around the following questions:

Why don't you get the same pattern every time?

Why do some of the patterns reach the edge and others do not? Why don't they all behave the same?

If we grew 1000 patterns, would you expect the fraction of patterns reaching the edge to be some definite number? Why or why not?

Would you expect the patterns to be different if we throw a single die and place a tree for numbers 1, 2, or 3 and a rock for numbers 4, 5, or 6? Would a greater or smaller fraction of patterns reach the edge than for the original game? Explain the reasoning behind your answer.

Suppose we throw a single die and place a tree for number 1 but a rock for numbers 2, 3, 4, 5, or 6? In this case would a greater or smaller fraction of patterns reach the edge?

Suppose we throw a die and place a tree for numbers 1, 2, 3, 4, or 5 and a rock for number 6 only. How will this change the pattern of the forest? Will it increase or decrease the chance that the forest reaches the edge?

Activity 3.2 Growing a Forest by Computer

Growing a forest on your grid by hand takes a lot of time. The computer program **Forest** grows forests much faster. Call up this program and grow a few forests with the probability 0.5, the chance that the next seed dropped will grow a tree.

Probability 0.5 comes from flipping a coin: equal chance for heads as for tails. But the computer can operate with different probabilities, for example reducing to only 0.1 the chance that the next seed dropped will grow a tree.

Take a Guess!

When the tree probability drops from 0.5 to 0.1, is the forest more likely or less likely to reach the edge? What about a tree probability of 0.9? Is growth to the edge more likely or less likely in this case? Suppose we grow many forests, each with a different tree probability, starting at 0.1 and increasing to 0.9. Will the fraction reaching the edge increase or decrease as tree probability increases? Will this increase or decrease take place gradually over different probabilities or suddenly at a particular tree probability?

Now check out your predictions by trying different tree probabilities, using the buttons across the bottom of the screen. How does the fraction reaching the edge change as you change the tree probability?

The computer can summarize your results. Click on the GRAPH button and look at the plot. Fill in the graph by clicking on probabilities you have not chosen before. At the bottom of the graph is the number of times you have tried each tree probability. Make several trials for any probabilities that interest you.

Do you see a pattern? How does the fraction of forests reaching the edge vary as the tree probability increases?

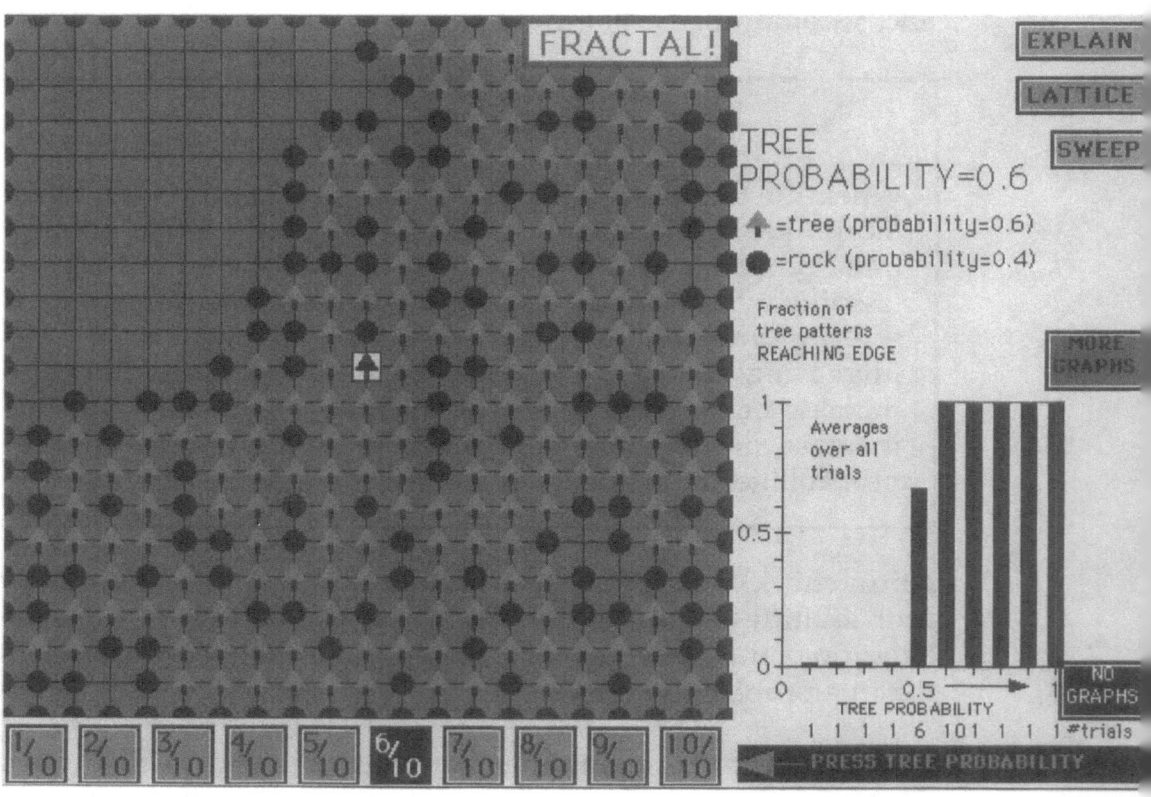

Explore!

The **Forest** program is intended to be self-explanatory. It should help you to explore percolation phenomena that frequently appear in Nature. Try out some of the following features, using them to explore percolation on your own:

- Different tree probabilities, from 0.1 to 1.0.
- Different kinds of lattice: square, triangle, honeycomb
- Different sizes of forest: 20 x 20 = 400 or 40 x 40 = 1600
- Interactive explanations
- Different summary graphs:
  - Fraction of patterns reaching the edge
  - Fraction of possible locations occupied by trees.
  - Number of trees inside circles of different radii

Think up your own questions and use the **Forest** program to explore possible answers. Here are some suggested questions

What do the different forest patterns have in common, whether small (20x20) or large (40x40) forests, square or triangle or honeycomb lattices? How are they different from one another?

Do you see evidence of a <u>critical value</u> for growth probabilities where forests seem to change from dying out to growing forever? How might we find that value?

Is the critical growth probability for a triangle grid (6 nearest neighbors) lower, higher, or equal to the critical growth probability for the square grid (4 nearest neighbors). Can you explain your result?

Do you expect that the critical probability for a honeycomb lattice (3 nearest neighbors) will be lower, higher, or equal to the critical probability for the square lattice (4 nearest neighbors). How about for the triangular lattice (6 nearest neighbors)? Use the button labeled LATTICE to explore these possibilities.

Have you done the Fractal Dimension unit? If so, what are the conditions for fractal forest patterns? What fractal dimension do you measure for the forest in these cases?

Report to the class on your investigations and conclusions.

Activity 3.3 Computer: An Adequate Model? Forest Size.

The Forest program does not show *real* forest growth. In fact, none of our computer simulations show real structures from Nature. Rather, the Forest program shows a *model* of forest growth -- actually several different models. How "good" are these models? How dependable are the conclusions we make as we use these models? In this section we provide instructions for investigating how results depend on the *size* of the forest. Begin by making a photocopy of the tables and graphs on two of the following pages.

Predict!

Think of a forest on a square lattice that is 4 x 4 in size: four rows of four trees each. Which of your observations on the larger forests will be true also for the 4 x 4 forest? Which results might be different?

Start up the **Forest** program and grow a few forests until buttons appear. Then select the small forest (20 x 20) and call up the loglog plot. (Click on **GRAPHS** button, then the **MORE GRAPHS** button, then the **LOGLOG** button.).

Ignore the graph at the right. Instead we are going to concentrate on the forest inside the different circles. Think of the first (inner) circle as the boundary of a tiny forest, the second circle as a slightly larger forest, and so forth. We compare results for different forests to see how these results depend on size.

Concentrate on whether or not the forest grows beyond the various circles drawn on the display. Does *at least some part* of the forest reach out past the first circle? past the first and second circle? past all circles?. Change the tree growth probability and observe the corresponding growth of the forest past circles of different radius. After playing around a while, fill in Table 1 for different probabilities.

> A single trial may give entries in several columns. For example, suppose some part of a particular forest grows past the first and second circles but

not as far as the third circle. Then in the row for that probability, you add one to the columns "Grows Beyond Circle 1" and "Grows Beyond Circle 2."

Table 3-1 Assembling Data on Forest Size

Probability	Grows Beyond Circle #1	Grows Beyond Circle #2	Grows Beyond Circle #3	Grows Beyond Circle #4	Total Trials
0.2					
0.3					
0.4					
0.5					
0.6					
0.7					
0.8					

Grow several forests with each probability. Some probabilities will lead to more interesting results than others—do more trials for the interesting probabilities. If several groups of students are carrying out this activity, combine your results in Table 3-2.

Table 3-1 Summarizing Data on Forest Size

Probability	Fraction That Grow Beyond Circle #1	Fraction That Grow Beyond Circle #2	Fraction That Grow Beyond Circle #3	Fraction That Grow Beyond Circle #4
0.2				
0.3				
0.4				
0.5				
0.6				
0.7				
0.8				

Now plot your results for each Circle—each column on one of the four graphs below. The result is a test of how growth patterns vary with forest size.

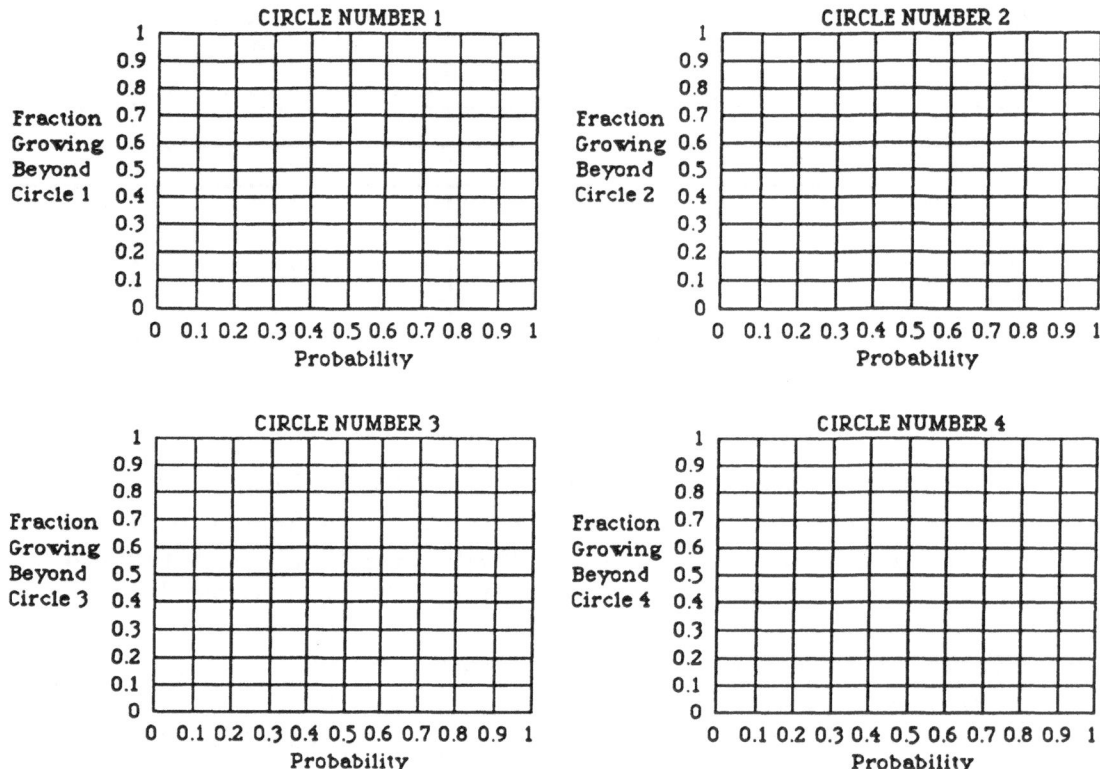

What seems to be happening for "larger forests," that is, forests bounded by larger and larger circles? What might the curve for a 100 by 100 grid look like? Is the evidence for a critical probability stronger or weaker for the larger forests?

So far we've created only a single cluster of connected trees, since each tree seeds only the sites next to it. In the following section we try another model of the forest by varying the rules a bit. We assume that the wind blows seeds everywhere; seeds can travel to <u>any site</u> not just those adjacent to a tree. This produces some isolated trees and small clusters of trees—in general a more realistic forest. Then we will see how easy or difficult it is to burn the forest down!

3.2 Burning a Forest

Activity 3.4 Burning a Forest by Hand

Now let's create a different kind of forest, then try to burn it down! To grow such a forest, work in pairs, each pair with a pencil, ruler, a paper of green self-stick dots, and a paper of red self-stick dots. Draw a 5 x 5 square grid on the paper, with each square approximately one inch on a side.

1. One student places his/her finger on the upper left square while the other student flips a coin.

2. If the coin is heads, place a green sticker (representing a tree) on the square. If it is tails, place nothing on the square.

3. Move the finger to the next square to the right.

4. Flip the coin and place a green sticker if the result is heads, nothing if it is tails.

5. Repeat this process until the coin has been flipped once for each square in the grid. Now we have made a forest.

Now burn the forest!

6. Place a red sticker over each green tree that lie in squares along the left edge of the forest. They represent a fire that starts on the left edge.

7. Look at each red sticker. Is there a green sticker-tree in the next square to the right? If so, place a red sticker over that green tree, burning it down.

8. Now look at each new red sticker. Is there a green sticker in the next square to the right, or above, or below it? If so, cover the green sticker with a red one.

9. Continue pasting red stickers over green trees next to burning trees (above, below, to the right, or to the left, but not diagonally) until no green tree is next to a burned one. Now you have burned your forest.

In the class as a whole, what fraction of the forests burned all the way from the left edge to at least one tree that lies along the right edge? What does this tell you about the connectivity of the forest? What would happen if the coin was lopsided ("loaded") so that it came up tails more than 50% of the time? Would there be more green stickers or fewer green stickers in your "forest"? Would the fire be more likely or less likely to spread across the resulting forest? What would happen if the coin was loaded in the other way, so that it came up heads more than 50% of the time? Would the fire be more or less likely to spread from one edge to the opposite edge?

Activity 3.5 Burning a Forest by Computer '

Now play **BlaZe!**, the computer game of forest management and fire fighting. Here is how the game works. You are responsible for growing and harvesting trees from a plot of land. The more trees you harvest, the higher your income. Unfortunately, your forest is located in a high-risk fire region; when the forest is grown, a blaze begins along the left edge. The more trees survive the fire, the more your profit. What tree probability should you use—the probability that a tree will grow on a given location? If you plant trees at low probability, trees will usually be separated from one another (low density), so the fire will not spread. As a result, not many trees will burn but you won't harvest many trees either, because there are not many trees altogether. On the other hand, if you use a high tree probability, the trees will typically be next to one another (high density), fire will spread across the forest, and again you won't have much of a harvest. The trick is to find the critical tree density (the critical probability) that gives you the chance for a large harvest but keeps the number of connections between clusters of trees low enough that you can control the spread of fire.

To help stop the fire, you fly a helicopter that drops water on the fire. The amount of water is limited, so you must find the "tree bridges" that connect big clumps of trees. Dropping water on these tree bridges is an efficient way to block the fire from spreading.

> **Guess:** How do you maximize your profit? The answer depends on the number of tree bridges. There are few tree bridges when there are few trees—low tree probability. Then you may not have to use the water at all: the fire just burns itself out in isolated clumps of trees. But few trees means low profit. So try the alternative: grow many trees by choosing a

high tree probability. But then there are many tree bridges—so many that you may not have time to dump water on all of them before the fire arrives. So most of the trees burn and your profit is low in this case too. So what is the trick to make a high profit? What probability should you use and why? (To encourage you to actually save trees, your score in the game is based on the number of <u>threatened</u> trees that you manage to save and the amount of water you used.)

What Do You Think?

Form groups to discuss the following questions. Be prepared to report your comments and conclusions to the class as a whole.

1. What is the difference between the forest grown in by the computer program Forest and the forest grown by the computer program BlaZe?

2. If a fire started somewhere in the forest grown by the Forest program and you were not allowed to use the helicopter to fight the fire, what fraction of the forest would burn down?

Would your answer to question 2 be the same or different if fire could start only along the left-hand edge of the potential growing region?

Which to you think is the more realistic model of a forest, the one shown in the Forest program or the one shown in the BlaZe program?

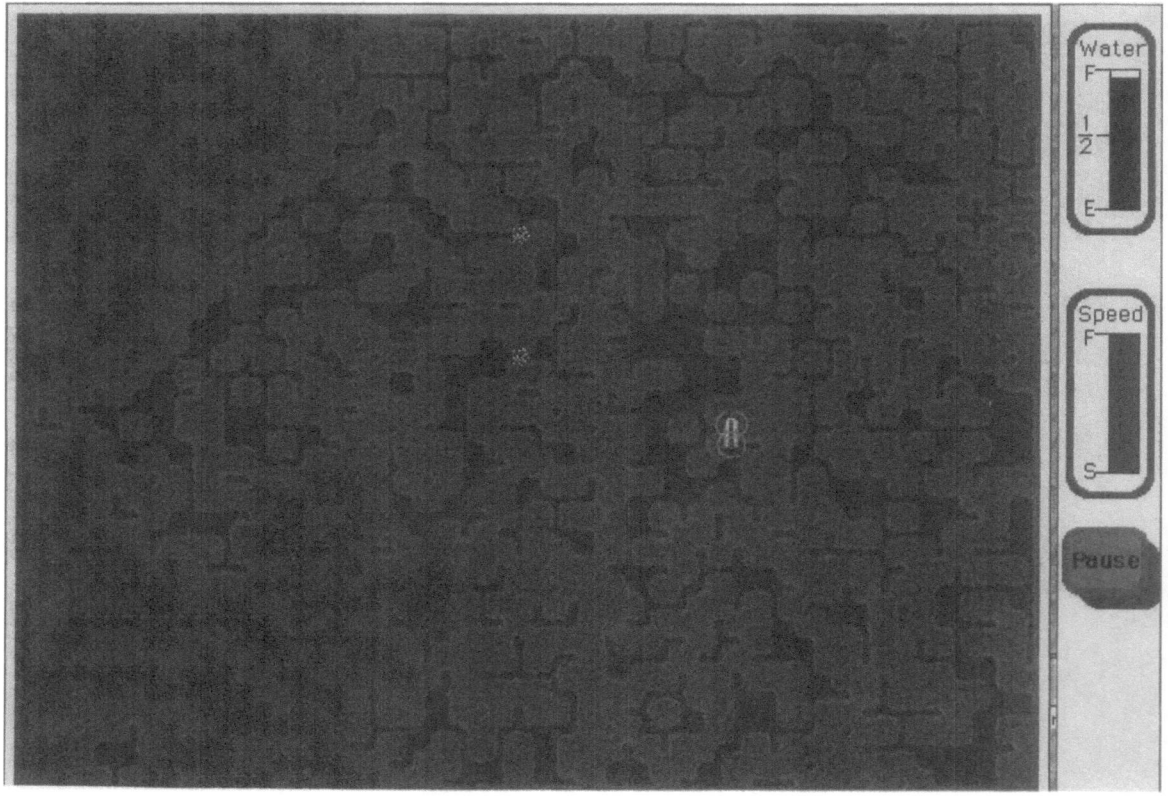

Activity 3.6 Jello Experiment

Forest fires are, strange to say, a very good model for gelation. Gels are made up of polymers, long-chain molecules, that attach to one another. If the chains are interconnected, spanning the sample, they can act as "bags" to contain the water, creating a gel instead of a liquid. In the same way, if clusters of trees span the forest, then there is a good chance that the fire will burn across the forest. In both cases the network spans the structure, connecting one side to the other. Thus, connectivity is the key idea for both forest fires and gels.

Gels are found in a wide range of systems, from the vitreous humor of the eye to the Jello on the dinner table. What gives Jello its peculiar properties? It is similar in some respects to liquids and in other respects to solids. What you have explored in the above activities should enable you to understand the characteristics of Jello, or gels in general. In particular, there is a critical concentration of gelatin molecules required to make Jello into a wiggly solid. Then the bag-compartments contain the water to keep the structure together. Below the critical concentration the bags do not span the structure, the water is not contained, and the substance flows as a liquid.

In the laboratory, experiment with different concentrations of gelatin to see if you can find the lowest concentration that leads to a gel rather than a liquid. How does raising the temperature affect your results?

Contents

Chapter 4. DNA and Literature

Chapter 4

DNA and Literature

There is no substance so important as DNA. Because it carries within its structure the hereditary information that determines the structures of proteins, it is the prime molecule of life. The instructions that direct cells to grow and divide are encoded by it; so are the messages that bring about the differentiation of fertilized eggs into the multitude of specialized cells that are necessary for the successful functioning of higher plants and animals. ...DNA has provided the basis for the evolutionary process that has generated the many millions of different life-forms that have occupied the earth since the first living organisms came into existence some 3 to 4 billion years ago.

-James D.Watson, Michael Gilman, Jan Witkowski, Mark Zoller, *Recombinant DNA*, Second Edition, Scientific American Books, New York, 1992, page 1

4.1 DNA Sequences

4.1.1 The Basics

Think of a rope ladder. A rope ladder is made of parallel ropes connected by rigid rungs. The rungs are the crosspieces on which you step to climb the ladder. The rope ladder can be twisted without changing its basic form.

The shape of a twisted rope ladder is like the famous "double helix" called DNA (Fig. 4.1). DNA resides in every cell of your body. It is a long chain of chemical building blocks that determines who you are genetically and what characteristics you can pass on to your children. Your body uses the information in DNA to build proteins, which are responsible for the structure, operation, and regulation of your body.

DNA is a *very* long rope ladder. How long? Each of the two strands of DNA has something like a hundred million rungs or crosspieces. Every single cell of your body (except sperm and eggs) has the same set of 23 different strands of DNA.

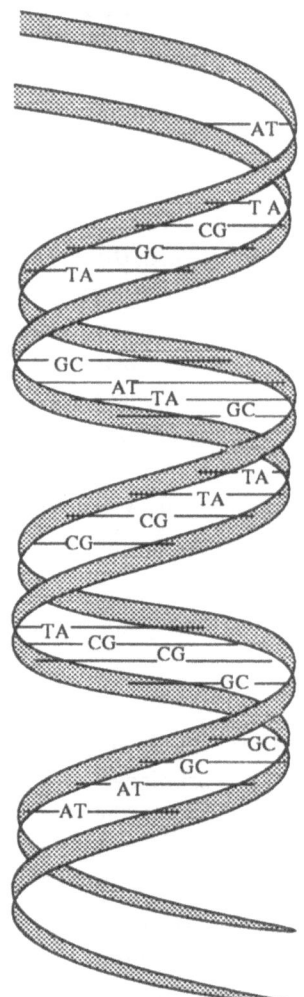

Figure 4.1: Sketch of a segment of a DNA molecule

◇ **Activity 1: Get a feeling for the length of a DNA molecule:**

- On a blank piece of paper draw a ladder from the bottom to the top of the page. Draw crosspieces on your ladder. Try to draw as many crosspieces as you can in this length of ladder.

- Now count the number of crosspieces on your length of ladder on this one page.

- Next, suppose you made many many photocopies of your page with the ladder on it. How many photocopies would you have to make so that the total number of crosspieces on all the pages would add up to 100 million? Assume 100 pages make a stack one centimeter tall. How tall a stack of paper would it take to contain 100 million crosspieces?

- Compare this height with the length of a football field—about 100 meters or 10,000 centimeters. [1] ◇

4.2 Storing Information in DNA

The information about your genetic makeup is stored in the DNA molecule. *How* is this information stored? In a computer, information is stored as "bits", which can be thought of as 0 or 1; only two possible choices. Computers calculate and do word processing exclusively with zeros and ones. Computer disks (CDs) store information in magnetized and unmagnetized regions, standing for one and zero. Compact disks store music and images in the form of pits burned into the plastic. A pit in a given region of the compact disk stands for the numeral one; the absence of a pit stands for zero. A kind of internal computer converts the string of 0's and 1's into music.

In DNA there are not just two choices, 0 and 1, but four choices for storing information, each choice represents a different building block. For historical reasons these four building blocks have fancy names whose first letters are A, G, C, and T. (See Box 1). They are called nucleotides (because DNA has been known to be a major constituent of the nucleus.[2]). On the DNA rope ladder, two of these nucleotides bond together in particular pairs to make each cross piece.

Actually, DNA would make an unsafe rope ladder, because each rung, each crosspiece, is weak in the middle. Every rung is made of either the combination A-T or the combination C-G (see Box 1). The rung easily breaks

---

[1] Answer to problem in Appendix A
[2] This does *not* mean, that DNA *only* occures in the nucleus!

BOX 1 WORDS! WORDS!

The study of DNA is full of fancy terms, terms that will be important if you decide to specialize in the subject.

DNA is short for **deoxyribonucleic acid** . Each rope of the DNA rope ladder is a row of sugars called **deoxyribose** linked together with phosphate groups. "Nucleic" is included in the name because DNA is a major constituent of the nucleus—not only - of the human cell.

The four building blocks with the first letters A, G, T, and C are called nucleotides: Adenine, Guanine, Thymine, and Cytosine. Because of their shape and chemical structure Adenine and Guanine belong to a group of bases called Purines, while Cytosine and Thymine belong to a group of bases called Pyrimidines. In brief, here is a classification scheme

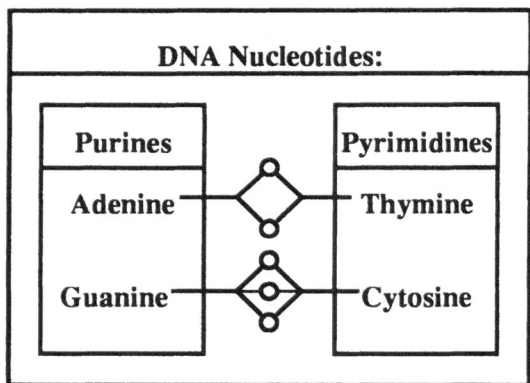

Why the separation between Purines and Pyrimidines? For one thing, each rung or crosspiece of DNA is made up of one Purine and one Pyrimidine: A bond connects A-T or G-C. These are called base pairs. Why one of each? Because the Purines are longer than the Pyrimidines; in order to have rungs of the right length, you need one of each. But why does Adenine always bond with Thymine (to give A-T) and Guanine always bond with Cytosine(to give G-C)? Why not the other way around? Because the bond in A-T is two hydrogen bonds, whereas the bond in G-C is three hydrogen bonds; you cannot join a two-bond A to a three-bond C or a two-bond T to a three-bond G.

It is the A-T bond or the G-C bond that breaks down the middle when DNA "unzips", leaving one nucleotide sticking to the "rope" spine on each side.

Of course, all these names mean nothing by themselves. They are labels, part of a larger language of chemistry and biology. But you can learn these labels and the details of base pairs that form the crosspieces in DNA and amaze your friends with your "mastery of the subject".

into two parts in the center, so that the rope ladder can "unzip" down the middle, leaving half of each rung sticking to the "rope" on one side. Each half-rung sticking out of the single rope makes up one of the four building blocks, A, G, C, and T, that carry your **genetic information**[3]. In fact, this is the way DNA makes copies of itself, by unzipping. Then each strand finds in the surrounding chemical soup the building blocks that reconstruct another rope ladder. In this way one rope ladder becomes two identical rope ladders; one strand of DNA becomes two identical strands of DNA(see Fig. 4.1).

The unzipped helix can reconstruct the full double helix. One side fits the other side the way the notches in a key fit the projections in a lock. Therefore it must be that the unzipped single helix carries all the information of the full double helix. We use this simplification in our project to study the unzipped helix, the "half rope ladder".

The sequence of letters A, G, C, and T form a kind of **alphabet**. Just like in an ordinary language, texts wirtten with this alphabet carry information. This information is read in our cells and used to build the different proteins that form our bodies and control their functioning. A section of a DNA sequence what carries the information to be translated into a certain protein is called a **gene** [4]

Table 4.1 lists 3,000 nucleotides in a piece of human DNA, part of a "tumor suppressor gene".

◇ **Activity 2: A Question**

- How many tables, each with the same number of entries as Table 1 would be required to list 100 million nucleotides in a human strand of DNA? ◇

4.2.1   Coding and Non-Coding Sequences in DNA

The sequence of A, G, C, and T nucleotides in a particular unzipped strand of DNA carries the information needed to create a rat or a chicken or a human. But workers have been astonished to find that most of this information appears to be wasted. Approximately 97% of the letter-sequences in human

---

[3]This is important because there are time when it its necessary for DNA to make exact copies of itself, as during the time right before a cell divides.

[4]You might have heard the term *gene* before. Today, it is a frequently used word. Expressions like "genetic disease, gene therapy, or gene technology" appear almost daily in the news. Or maybe, you heard people saying "it is in my genes". Actually, the term *gene* is quite old. It was created at the beginning of the century from a Latin word and means something like "hereditary factor" because genetics started as research on heredity. Today we in fact know that "our genes" which for instance determine the colors of our eyes or hair are passed to us by our parents.

1	ggatccagat	tcttttgaaa	ttcctcctgc	accaatatca	gcatttctac	cttctctgta
61	ggttggcttg	cctcacgtta	caatggctgc	agcaatcaga	agtgtcacat	cctcagtaat
121	acttaatatt	attactatta	tttaataata	tttagaactg	tgccatccct	gtttcaattt
181	atcaggctcc	cagcagacta	ctccttatct	ttcaaatgtc	aaaactgcat	cctgagctct
241	tgcctaaact	aatctggggt	gaggtgaatg	gaagtagcac	tttaattgta	ttcattcttt
301	gtagctggac	ctgggcctgg	gctatctcct	gacatttgcc	cacaagaaag	atttctgaag
361	ttaggtagga	atggctgttg	agtaggccag	tgcttgccaa	acctttacac	atcctcacat
421	atgtcataat	atgcagataa	aaagataatc	ccttatacaa	cttgctggga	taaactcagg
481	aggcttacag	catgacctgc	ttgaaggttc	ttcctgcctt	agaccttgct	cagctgctcc
541	aggatgaggg	gatttacatc	acagcacaac	tgtattttat	tcacagcata	aaccatctct
601	ttccttctca	gttgacgagt	tcagatgggc	aataacagtg	tctgccaaag	agaaaaaaaa
661	atgtattcaa	actagataat	ctattggtac	aaataccgag	acacagaagt	gataacagct
721	ttaagccaat	gtttgatggt	ggtagtccca	gcaagctctt	ttctgatgtc	tttgtgcctt
781	tgcacatgct	ccttctctgt	cactgttttc	ttcatcaaac	ataatataat	ggacaagtgg
841	aatcaaatag	aattgagttc	aaattctctg	ctacccatcg	gccctggtat	tggacaaatt
901	aactcctctg	agcctgtttc	ctcatctgca	acgtagacta	gctaatacta	cccattggaa
961	agcgttgttt	cttagctaat	gcatgcaagg	cttaaaacct	agatgacggg	ttgataggtg
1021	cagcaaacct	ccatggcata	cgtatgccta	tgtaacaaac	ctacacgttc	tgcacttgta
1081	tcccggaact	taaagtaaaa	aaaaaaaaaa	aaaaaaaaag	aaagaaagaa	aaagaaaaaa
1141	aaggctgttt	ctggggatta	aataagacaa	ttatgtaagg	tggccagcac	agttcctggt
1201	acatagtaaa	tgtcaggcct	gcctgacaga	cttctattca	gcagctactg	ctcccctgaa
1261	aatcttcctc	agacgtttcc	acggtgcttc	ccgttcttac	accactacaa	tcctttatta
1321	cactactatc	cgttcattcc	ccacagctcc	ctcccttcct	ttccctaacc	agtgatccca
1381	aaaggccagc	aagtgtctaa	cattttctat	cttctaagtg	actggtaaag	ttccgcacct
1441	atcagcgctc	caagtttgtt	tttgttttgg	ccgactttgc	aaaacggatt	gggcgggatg
1501	agaggtgggg	ggcgccgcca	aggagggaga	gtggcgctcc	cgccgagggt	gcactagcca
1561	gatattccct	gcggggcccg	agagtcttcc	ctatcagacc	ccgggatagg	gatgaggccc
1621	acagtcaccc	accagactct	ttgtatagcc	ccgttaagtg	caccccggcc	tggaggggt
1681	ggttctgggt	agaagcacgt	ccgggccgcg	ccggatgcct	cctggaaggc	gcctggaccc
1741	acgccaggtt	tcccagttta	attcctcatg	acttagcgtc	ccagcccgcg	caccgaccag
1801	cgccccagtt	ccccacagac	gccggcgggc	ccgggagcct	cgcggacgtg	acgccgcggg
1861	cggaagtgac	gttttcccgc	ggttggacgc	ggcgctcagt	tgccgggcgg	gggagggcgc
1921	gtccggtttt	tctcagggga	cgttgaaatt	attttttgtaa	cgggagtcgg	gagaggacgg
1981	ggcgtgcccc	gacgtgcgcg	cgcgtcgtcc	tccccggcgc	tcctccacag	ctcgctggct
2041	cccgccgcgg	aaaggcgtca	tgccgcccaa	aaccccccga	aaaacggccg	ccaccgccgc
2101	cgctgccgcc	gcggaacccc	cggcaccgcc	gccgccgccc	cctcctgagg	aggacccaga
2161	gcaggacagc	ggcccggagg	acctgcctct	cgtcaggtga	gcgagcagag	ccgccgtcgc
2221	ctcacgcggg	aagggcgccc	cgggtgtgcg	tagggcgggc	gcaaggcggc	tcggcgggga
2281	cccgtcctcg	ccaggggccg	ggtcccggcg	ggaggaggcg	ccctccctgc	cccccgccac
2341	ggcggagcgt	ctgcagaatg	gtgacaggat	tctgggttct	tgggcgaggg	gtctcggctt
2401	caacttgaca	ggtgtcgggc	gggtggggct	agggtcctga	gcgaagtgac	aggtgcagtt
2461	ccctcttgtg	aggctcggag	gcagagggtc	gttgcgagcg	tccatcagac	gcaaaaaatg
2521	aaaaataaaa	atacaaaaat	ggtgtctgtg	ggagagtttt	tcaccggaga	attggagtac
2581	tccggtggtc	gtctgacttt	ctgttttggt	tcacgcgatg	caacagttgg	gaagtatttt
2641	cttccgggcg	tgcactgcat	ctgaagtcca	tttgtgggag	aggccgacca	gaaagccttg
2701	gacaagaagc	gcagggtcct	gagtgtccat	tgcccacagg	atactcggct	caggagcttt
2761	gcggcgtttc	cttagaacaa	taatgcatcg	aggccttggg	gactcaaagc	catctgtagt
2821	gattgatgga	gcgtaactct	ttagaggaac	tgaaacatgg	gcaaaacttt	catgagacat
2881	ttaccagaag	tgcttgaaag	tttctaaact	ttttttttttc	ctgtttgatg	aactcttctt
2941	gcgtgttagt	cggcttcggc	ttgtctcatt	atttcttcca	ttttgccttt	tgactttgaa

Table 4.1: A sequence of DNA from a "tumor suppressor" gene. The number at the beginning of each line is the position of the nucleotide that follows next.

BOX 2 Why "junk" sequences in DNA?

"The protein-coding portions of the genes account for only about 3% of the DNA in the human genome; the other 97% encodes no proteins. Most of this enormous, silent genetic majority has long been thought to have no real function—hence its name 'junk DNA.' But one worker's trash is another worker's treasure, and a growing number of workers believe that hidden in the junk DNA are intellectual riches that will lead to a better understanding of diseases (possibly including cancer), normal genome repair and regulation, and perhaps even the evolution of multicellular organisms." Rachel Nowak, "Mining Treasures from 'Junk DNA' ", Science, Vol 263, 4 February 1994, pages 608-610.

DNA are cut out and thrown away during construction of proteins. The protein is built using only the remaining 3%- the **coding sequences**. The unused portions are called the **non-coding sequences** - sometimes called junk sequences. Even a single gene can contain non-coding parts. In that case we call the coding parts **exons** and the non-coding parts **introns**. For this discovery, the 1993 Nobel Prize in medicine was awarded to the Americans Richard Roberts and Phillip Sharp. Look again at Table 4.1. Parts of the series there are non-coding or "junk" DNA. Can you guess which parts are "junk"?

Why is most of the DNA code discarded? Are our bodies really so inefficient in using genetic information? Are non-coding sequences just left-overs, chunks of "old instructions", perhaps useful at an earlier stage of evolution but not needed any more? Or are they really "junk"? Could it be that these non-coding sequences carry information useful in some way other than coding for protein? Nobody knows the answers to these questions yet (Box 2). This is an area of current research being carried out right now. And some of the research methods use skills we have already developed in earlier units. Our goal in this project is to use these skills to investigate coding and non-coding sequences of DNA.

4.2.2   Measures of Information

The sequence of the four nucleotides, A, G, T, C forms a kind of language that carries genetic information in DNA. A lot of people are trying to translate this language—to discover the information it contains. We know information is present, at least in the coding sequences of DNA. How do we know? Because these coding sequences are used in creating proteins (p. 159). And we have some knowledge about the intermediate steps by which sequences of A, G, T, C are used to construct components of protein. Still, there is still much to learn about the structural and control purposes of many a long sequence in human DNA.

Do the non-coding ("junk") sequences of DNA also carry information? A lot of people are trying to find this out also. You can take part in this effort by working through this workbook. But how can we possibly discover the answer—how can we look for information in the "junk" sequences—when we do not yet have a complete translation of even the coding sequences of DNA?

Is there some method for measuring the "amount of information" in a language without actually *knowing* the language? without being able to *read* the language? Answer: There is no sure-fire way to measure information in an unknown language, but there are at least two methods that give some evidence of the presence of information. One is a study of **word occurrences** in the language. The other involves turning the language into a **visual landscape**. In this project we use both methods, starting with word occurrences.

4.3   Word Occurrences and Zipf's Law in Literature and DNA

4.3.1   Word Occurrences and Zipf's Law in Literature

Suppose you want to make a dictionary for young students, a dictionary short enough, light enough, to be easily carried around. What are examples of words you would include in your dictionary and which words would you leave out? One way to decide is to take a large sample of different kinds of literature that students may read and count the number of times that different words appear in this writing. Put these words in an ordered list, with the most-used words first, then include in your dictionary the first 5,000 words in this list (for example).

◇ **Activity 3: Word Count**
Make a list of the occurrences of the words in the following paragraph:

- On a seperate piece of paper write down a list of the words, starting with the first.

- When a word occurs for a second time, do not add it to the bottom of the list, but rather put a mark next to that word on the list. Add more marks as words occur again and again.

Table 4.2: Word occurrences in a text paragraph. Activity

	word	occur-rences		word	occur-rences		word	occur-rences
1			11			21		
2			12			22		
3			13			23		
4			14			24		
5			15			25		
6			16			26		
7			17			27		
8			18			28		
9			19			29		
10			20			30		

- After completing tabulating all words in the paragraph, rearrange your list in order, with the most-used word first and the least-used word last. Write your results for the first 30 words into Table 4.2.

Ancient DNA

Excerpts from an article by Svante Pääbo in *Scientific American*, November 1993

Genetic information that had seemed lost forever turns out to linger in the remains of long-dead plants and animals. Evolutionary change can at last be observed directly.

Most of our knowledge of the molecular processes that underlie evolutionary change is based on the comparison of the genes of living species. From such differences, molecular evolutionists infer the historical changes that gave rise to presentday DNA sequences. Yet these studies are tentative in nature. Unlike the remains of animals and plants, DNA molecules do not leave impressions in rock. Biologists therefore despaired of ever being able to check their conclusions against the historical record, as paleontologists do.

But in the past decade workers have learned that ancient DNA, though degraded, sometimes survives the ravages of time, and molecular biologists have perfected methods of amplifying these trace amounts of ancient DNA. Worker have so far used DNA from bone and soft tissues to establish reliable sequences for seven extinct mammals. The oldest was the woolly mammoth—a frozen carcass that was found in the permafrost of Siberia. More such studies are under way, including efforts to decode DNA extracted from insect entombed in amber millions of years ago. We can thus look forward to learning much more about the genetic relations among extinct species.

Table 4.3: Table of word occurrences for a long text. Activity

Text source:

word	occur-rences	word	occur-rences	word	occur-rences
1		17		33	
2		18		34	
3		19		35	
4		20		36	
5		21		37	
6		22		38	
7		23		39	
8		24		40	
9		25		41	
10		26		42	
11		27		43	
12		28		44	
13		29		45	
14		30		46	
15		31		47	
16		32		48	

- Now, load a longer text into your favorite word processor and count word occurrences.

- Start with the first word and count its numbers of occurrences. Try the second one, etc

- Start with the word that appears first in the text, and then try the second woard in the text, and so on.[5] Write your results for the first 48 words and the name of the text source into Table 4.3. ◇

Table 4.4 shows the first 180 entries of a word list compiled from 5,088,721 words of children's literature. The word with the largest number of occurrences was "the", which appeared 373,123 times in this large sample. The word with the second largest number of occurrences was "of", which appeared 146,001 times, and so forth. The 180th word is "under" with 2989 occurrences. The table does not show the total of 86,741 different words found in this sample, more than 1,700 of which each occur only once.

---

[5]Does your word processor have a function which helps you counting words? Otherwise use a "find" option and count. We know, it is a tedious job. Fortunately, for further activities you will be using a program that does the counting and rearranging automatically.

The 5,088,721 words were collected in 500-word samples from 1,045 different published texts with the purpose of providing a word list for a dictionary appropriate for students in grades 3 through 9. In making such a list, one must decide what is meant by "a word". In this collection, different variations of the same word (including capitalized variants) are treated as different words. For example, *word*, *Word*, *worded*, *wording*, *words*, and *Words* are each treated as a different word.

Table 4.5 shows a more "grown up" list of words, the number of occurrences of different words in a sample of 46,201 words taken from a selection of articles in an electronic encyclopedia (The Compton Encyclopedia). Here the choices were made somewhat differently: A capitalized and uncapitalized word (such as Word and word) were treated as the same word. However, any sequence of letters followed by a period was defined as a word. For example U. S. A. would lead to three listings, under the separate single letters u and s and a.

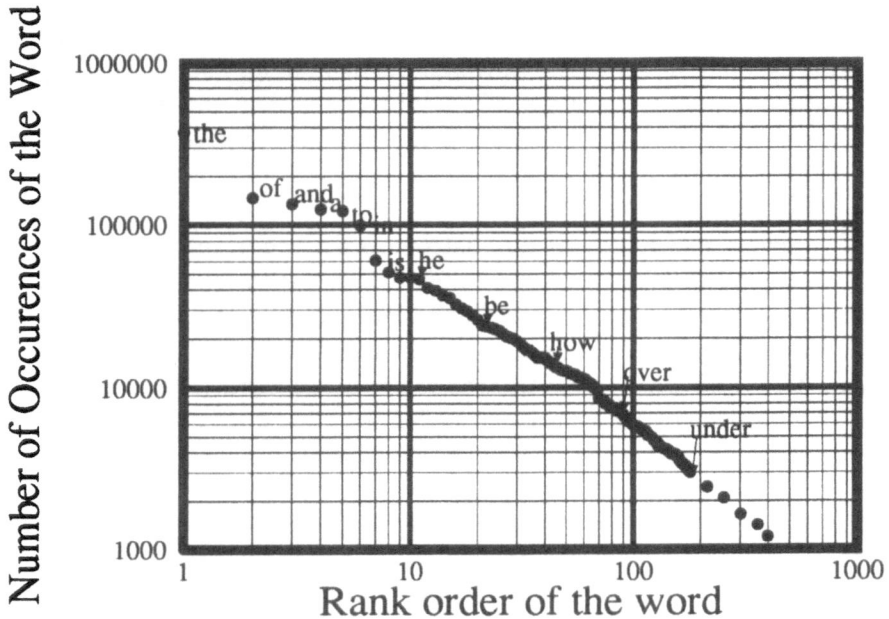

Figure 4.2: Plot of word occurrences from Table 4.4; about 5 million words of articles for children. A few selected words are added at the lower right to show the continued trend of the slope.

George Kingsley Zipf (1902-1950) discovered that when one plots these kinds of tables on a *log-log graph*, one gets a more or less straight line. The graphs for these two tables are shown on the double-page spread following the tables (page 170-173). In these graphs the vertical axis measures the number of occurrences of each word and the horizontal axis measures the order of the word: first, second, third in the number of occurrences - we call this kind of graph a **rank plot**. Box 3 describes the kind of graph: number of occurrences of the word on the vertical axis and word rank (that is, the numerical position of the word in the occurrence table) on the horizontal axis.

◇ **Activity 4: Rank plots.**

- Plot your results from Table 4.2 and Table 4.3 in a rank plot on log-log-paper. You can use the log-log paper in appendix B. Do you obtain straight lines[6]?◇

---

[6]If not, it could be that your text is just too short. Usually you need texts with a total number of ~ 40,000 words. See also box 4.

Table 4.4: Occurrences of words in 5,088,721 words of childrens' reading

Word	Occur-rences	Word	Occur-rences	Word	Occur-rences	Word	Occur-rences
1 the	373123	46 their	13258	91 may	6635	136 even	4225
2 of	146001	47 if	12907	92 way	6612	137 such	4223
3 and	133899	48 up	12776	93 only	6583	138 because	4207
4 a	124959	49 do	12695	94 long	6220	139 here	4184
5 to	121347	50 will	12646	95 little	6204	140 why	4147
6 in	99108	51 about	12496	96 me	6180	141 went	4132
7 is	60852	52 out	12252	97 number	6059	142 take	4089
8 you	50957	53 many	12158	98 very	5997	143 things	4070
9 that	47443	54 then	12022	99 after	5915	144 men	4067
10 it	47284	55 them	11997	100 back	5862	145 years	3966
11 he	46249	56 these	11611	101 just	5858	146 put	3942
12 was	40934	57 so	11543	102 called	5789	147 us	3929
13 for	39322	58 some	11534	103 most	5785	148 different	3926
14 on	36482	59 her	11375	104 our	5777	149 say	3916
15 are	35454	60 words	11215	105 get	5700	150 old	3894
16 as	32208	61 would	11188	106 know	5655	151 again	3892
17 with	30455	62 other	10729	107 where	5611	152 help	3875
18 his	29268	63 him	10703	108 used	5607	153 off	3873
19 they	27620	64 into	10620	109 man	5486	154 great	3855
20 I	25932	65 has	10369	110 new	5448	155 away	3814
21 at	23975	66 two	10085	111 through	5442	156 name	3766
22 be	23746	67 more	9992	112 go	5388	157 Mr	3748
23 this	23301	68 write	9846	113 much	5386	158 tell	3715
24 from	22799	69 like	9696	114 good	5343	159 air	3673
25 have	22337	70 could	8585	115 before	5275	160 set	3572
26 or	21283	71 see	8518	116 too	5071	161 small	3555
27 had	20511	72 no	8483	117 any	5023	162 big	3476
28 by	20189	73 time	8441	118 same	5022	163 should	3470
29 one	19976	74 make	8333	119 day	5019	164 still	3421
30 but	19196	75 people	7989	120 look	4933	165 every	3398
31 not	18645	76 than	7982	121 came	4914	166 give	3366
32 what	17709	77 my	7898	122 right	4815	167 found	3362
33 were	17031	78 first	7655	123 think	4746	168 between	3324
34 all	16997	79 been	7645	124 come	4676	169 home	3308
35 we	16452	80 who	7576	125 sound	4667	170 line	3293
36 when	15886	81 word	7532	126 also	4647	171 below	3276
37 your	15311	82 its	7512	127 around	4632	172 1	3222
38 said	15309	83 now	7457	128 three	4413	173 looked	3197
39 can	15247	84 down	7206	129 does	4408	174 sentence	3122
40 there	15194	85 water	7194	130 another	4377	175 never	3115
41 an	14696	86 did	7169	131 work	4358	176 read	3057
42 each	14290	87 made	7073	132 must	4307	177 last	3030
43 which	14016	88 use	7009	133 part	4285	178 2	3030
44 she	13653	89 find	6916	134 well	4255	179 own	3006
45 how	13303	90 over	6882	135 place	4240	180 under	2987

Table 4.5: Occurrences of words in 46,201 words in a collection of articles from the Compton Encyclopedia. Courtesy of Rosario Mantegna.

Word	Occurrences	Word	Occurrences	Word	Occurrences	Word	Occurrences
1 the	2375	46 most	53	91 some	29	136 sides	22
2 of	1167	47 first	52	92 during	29	137 institutions	22
3 and	1131	48 be	52	93 been	29	138 order	22
4 in	875	49 also	51	94 n	29	139 power	22
5 to	512	50 banks	50	95 c	28	140 shares	21
6 a	412	51 into	49	96 between	28	141 central	21
7 by	271	52 under	48	97 until	28	142 court	21
8 was	230	53 major	48	98 through	27	143 there	21
9 is	217	54 may	46	99 union	27	144 trade	21
10 or	189	55 indian	46	100 west	26	145 german	21
11 for	170	56 indians	46	101 such	26	146 body	21
12 on	170	57 system	42	102 english	26	147 blood	21
13 as	147	58 government	42	103 civil	26	148 loans	21
14 with	147	59 became	42	104 can	26	149 among	20
15 that	141	60 many	42	105 center	26	150 sq	20
16 from	140	61 city	41	106 than	26	151 side	20
17 are	135	62 banking	41	107 b	26	152 stocks	20
18 s	125	63 york	41	108 market	26	153 economic	20
19 it	114	64 have	41	109 president	26	154 developed	20
20 u	106	65 one	40	110 against	25	155 company	20
21 stock	105	66 not	39	111 established	25	156 chicago	20
22 he	102	67 amendment	38	112 celestial	25	157 bible	20
23 his	101	68 federal	38	113 this	25	158 years	20
24 see	92	69 g	38	114 however	25	159 nation	20
25 right	92	70 more	38	115 include	25	160 national	20
26 war	91	71 idea	38	116 i	25	161 military	20
27 an	89	72 early	36	117 language	25	162 important	20
28 american	88	73 great	36	118 law	25	163 although	19
29 were	86	74 ii	36	119 about	24	164 average	19
30 cent	83	75 america	35	120 abortion	24	165 any	19
31 which	75	76 states	35	121 angle	24	166 act	19
32 but	74	77 state	34	122 spoken	24	167 where	19
33 new	73	78 constitution	34	123 capital	24	168 economy	19
9 34 other	72	79 financial	34	124 called	24	169 development	19
35 they	71	80 had	34	125 congress	24	170 commercial	19
36 languages	67	81 south	33	126 charles	24	171 republic	19
37 world	65	82 john	33	127 france	24	172 form	19
38 its	65	83 led	33	128 today	24	173 french	19
39 at	62	84 angola	32	129 only	24	174 troops	19
40 after	60	85 who	32	130 parliament	24	175 began	19
41 all	58	86 two	32	131 w	23	176 history	19
42 e	55	87 when	31	132 group	23	177 made	19
43 their	55	88 germany	31	133 known	23	178 island	19
44 has	54	89 bank	31	134 left	23	179 largest	19
45 north	54	90 exchange	30	135 spanish	22	180 later	19

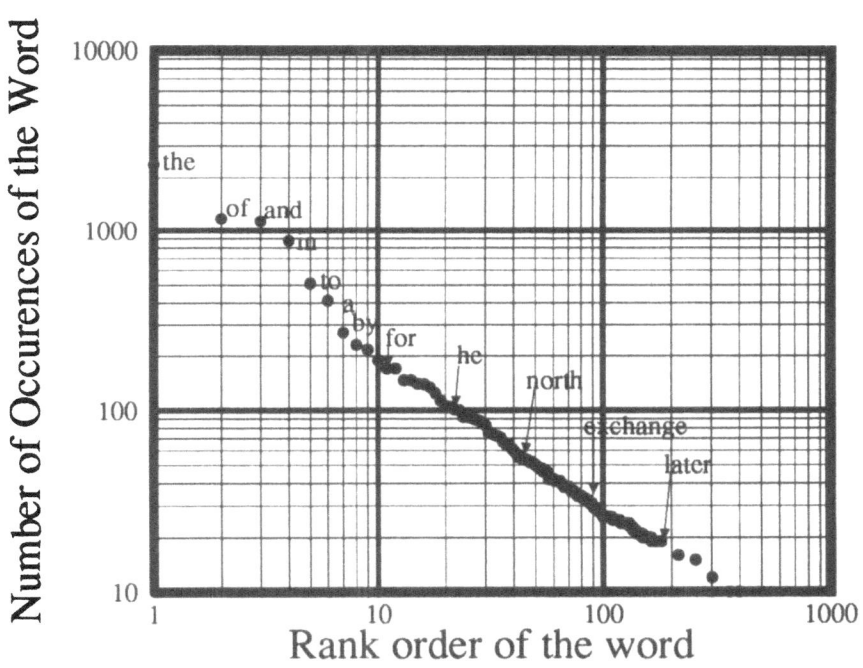

Figure 4.3: Plot of word occurrences from Table 4.5; articles from the Compton Encyclopedia. A few selected words are added at the lower right.

◇ Activity 5: Wordcount

- Using the computer program **Zipf** make an occurrence table for a given long text from one of the following sources[7] ,

    – a CD-ROM disk of Time magazine issues (text only).
    – Dr. Seuss, the author of simplified-English children's books
    – James Joyce from the BU project.
    – all of the Bible
    – selected books of the Bible.

- Plot the resulting table on log-log paper and measure the slope. Write down your results in Table 4.6

- Try out books from different categories: romantic novels, Mark Twain, the philosopher Bertrand Russell

- Compare the slopes. Do you observe any trends?

- Compare the results with results of your classmates.

- Do you have a text of your own on diskette? **Try it!** Does your text give a smooth curve?

- Try also books in a foreign language. Write your results in the table. What do you conclude?◇

You have now analyzed different books, maybe even in different languages. Probably you found that by making a rank plot you gain a straight line of slope close to 1.0. Isn't it amazing that you find this behavior independent on language and author? Maybe the slope changes from author to author, but the overall behavior remains the same. Natural languages show this kind of behavior. In the next section we will analyze DNA sequences applying the techniques which we introduced for analyzing literature. Can we find differences in coding and non-coding regions?

---

[7] Your teacher knows which of those are available

BOX 3 Rank plots and Zipf's Law

In order to construct a rank plot we take a set of samples and measure a certain quantity, e.g. we could measure the heights of your classmates, the population in big cities, the number of word occurrences in a text, etc. After having collected all our data we can compare all the numbers we found, and order them: the largest number will be the first, the second largest the second, the smallest number will be the last one. We can assign a *rank* to these numbers: The largest number has rank 1, the second largest number rank 2, the third rank 3, etc. When we plot now our numbers versus their ranks, we call this graph a **rank plot**. *Zipf* found, that when we plot the number of word occurrences versus their rank on log-log paper, we get a straight line (more or less). We call this finding **Zipf's Law**.

Table 4.6: Slopes for different books in a Zipf plot. Activity

books	author	slope
1		
2		
3		
4		
5		
6		
7		
8		
9		
10		

BOX 4 Zipf's Law, Fractals, Power Laws, and Finite Size Effects

Caution: This Box contains mathematics—Viewer discretion is advised. Perhaps you have worked already with the **Fractal Coastline** or the **Fractal Dimension** programs. Then you are already accustomed to log-log plots and you know what a straight line in log-log plot means. Maybe you want now to learn how the Zipf law is connected to fractals. The answer is: They both belong to the so called "scale free" phenomena. That means the following: If you have a geometrical fractal (e.g. the Sierpinski Gasket) you take a part of it and blow it up to the size of the original fractal, then the portion looks exactly like the original fractal, i.e. on any length scale the fractal looks the same. A consequence of this is that the number of boxes of size l which we need to cover a fractal, follows a power law: $N(l) \sim l^{-D}$ where D is the dimension of the fractal structure. If we consider boxes of two different sizes e.g. l_1 and $l_2 = 2l_1$, then $N(l_2)/N(l_1)$ is always equal 2^{-D} independent on the absolute size of l_1 and l_2.[a]. We observe the same phenomenon in the Zipf law: Consider two words w_1 and w_2 with ranks r_1 and $r_2 = 2r_1$. It does not matter how large r_1 and r_2 really are, the number of occurrences of w_2 divided by the number of occurrences of w_1 will always be approximately the same: 2^{-m} where m is the slope of the straight line in the rank plot. However, the straight line is not infinitely long and after a while it falls off when it gets to the rare words. This is a consequence of the finite size of the text. If you take a text which is, say, ten times as long as the text you have analyzed, then the straight line will continue straight beyond where it fell off before, until it reaches another set of rare words. For an *infinitely long* text[b] the straight line would go on for ever. Therefore physicists like to call this kind of behavior **finite size effects**.

---

[a]That means, it does not matter if $l_1 = 2in$ and $l_2 = 4in$ or if $l_1 = 0.3in$ and $l_2 = 0.6in$, in both cases $N(l_2) = N(l_1)2^{-D}$

[b]⌣Physicists like to talk about infinite structures; however, none of them ever saw one. ⌣

4.3.2 Word Occurrences in DNA

In the following section we will study DNA using the concept of "word counting". But before we start to analyze DNA sequences we have to solve a **big problem:** Imagine the following: You find a book or a text which is written in a foreign language which you do not know. You could for instance find the following text in a song book[8]:

Nell

Ta rose de pourpre à ton clair soleil,
O Juin, étincelle enivrée,
Penche aussi vers moi ta coupe dorée:
Mon cœur á ta rose est pareil.
Sous le mol abri de la feuille ombreuse
Monte soupir de volupté;
Plus d'un ramier chante au bois écarté
O mon cœur, sa plainte amoureuse.
Que ta perle est douce au ciel enflammé,
Etoile de la nuit pensive!
Mais combine plus douce est la clarté vive
Qui rayonne en mon cœur, en mon cœur charmé!
La chantante mer, le long du rivage,
Taira son murmure éternel, Avant qu'en mon cœur, chère amour,
ô Nell, ne fleurisse plus ton image![a]

Text by Leconte de Lisle

---

[a]Translation in app. C

Perhaps you will recognize the language as French, but suppose you do not understand the text of the song. Still you can count the words in the text and make a rank plot (though in this case that would not make much sense because the text is much too short!). Anyway, the spaces between words and the punctuation in the text help you to identify single words. A DNA sequence does not offer us this kind of luxury. We do not know the "spaces" in a DNA sequence. We do not even know if there are any. To illustrate the problem, try out the following activity.

◇ **Activity 6: Spaces**

• Take your text editor on the Macintosh and write a letter to a friend in

---

[8]If you *do* understand French, please assume for a moment, that you do not!

the classroom. (Or if you like, to your teacher) Use only capital letters or lower case letters.

- Take out all the spaces in the text.

- Give this letter to your friend to read.◇

Probably your friend will still be able to read the letter because he or she knows the underlying language and knows where to start reading each word. You could begin and end your letter with some randomly chosen letters. Your friend will still be able to read the text as soon as he/she tries to find the first meaningful word, i.e. combination of letters.

This is our problem in DNA: We do not know the underlying DNA language[9]. Thus we have to ask if there still exists a way to apply the *Zipf method* to DNA sequences.

4.3.3   n-Letter Words in Literature

We approach the problem by defining "a word" in a sequence of letters without spaces as a group of n letters where n can be 2, 3, 4, 5, 6, ... Look again at the letter which you have written to your friend. It is a sequence of many letters without spaces. Now you group these letters into, say, 4-letter words. The first 4 letters make up the first word, letter 5-8 make up the second word, and so on. In a very long text we can now again count the number of occurrences of these 4-letter words. The question is, if this definition of a word really works. In order to find out if the concept of n-letter words is reasonable, we first look again at the literature we have already analyzed, because there we know what happens for real words. So we have always a good possibility for comparison.

◇ **Activity 7: n-letter words in literature**

- Choose three of the books that you analyzed in activity 5.

- Have the program eliminate spaces.

- Analyze the result again using either 4-, 5-, or 6-letter words[10].

---

[9]Although we know in principal how the information for the proteins is encoded in the DNA sequence(see later Box 5), we still do not know where we are supposed to start reading a coding section in the DNA sequence. Moreover, exons have not necessarily to start with a start codon or to stop with a stop codon! The stop codon will be the *last* codon in the *last* exon of a gene.

[10]We recommend that for one group analyses a text with 4-letter words, another group analyzes the same text with 5-letter words. Thus you can gather more data

- Determine the number of occurrences of n-letter words with the program **Zipf.**

- Let the program construct a rank plot and determine the slope of the curve.

- Write the result for the slope into Table 4.7

Table 4.7: n-letter words. Slopes of Zipf plots. Activity

books	n	author	slope
1			
2			
3			

- Compare the new results with your previous ones.

- Do not turn the page until you have completed the comparison!!◇

Probably you got again a straight line in your rank plots, but you found the slope (~ 0.5) to be significantly smaller than in your first graphs. But still the rank plot also for n-letter words seems to give useful results, because you still get a straight line. We will now use the rank plot to find information about DNA sequences.

4.3.4   n-Letter Words in DNA

In this section we generate rank plots for DNA sequences. But first we may consider the total number of different 2-, 3-, 4-,...letter words (n-letter words in DNA are groups of n nucleotides) which we actually can find in a nucleotide sequence. We have to remember that in a DNA sequence there can occur four different bases which we call by the first letter of their names: **A**, **C**, **G**, and **T**. **Question:** How many 1-letter words can we build from these letters (nucleotides)? **Answer:** If $n = 1$, you have just four possibilities: the letters[11] themselves *A, C, G, and T*.

$\diamond$ **Activity 8:   How many different 2-letter words can we find in a DNA sequence?**

- Write down all the possible combinations of 2-letter words you can find using just the letters A, C, G, and T[12].$\diamond$

Solutions:

_____

_____

_____

_____

In the next activity we will try to find the total number of all possible 3-letter words.

$\diamond$ **Activity 9:   3-letter words in DNA**

- Divide your class into four groups.

- One member of each group writes down all possible three letter DNA "words" that begin with *A*, a second all the three letter words that begin with *C*. The other two members of each group write down the three letter words that begin with *G* and *T*.

---

[11] In the following we will use the words "letter" and "nucleotide" interchangeably.

[12] You can compare your results with the solution in the appendix A

Figure 4.4: (p. 182) Tree structure to determine all possible 3-letter DNA words. Reading starts at the left side. The letter symbol at the left side represents the first letter. Each letter is connected by lines with four letters representing the second letter in the DNA word. To determine the number of 2-letter words you have to count the total number of lines connecting the first with the second letter. Each second letter is connected by lines with four letter symbols representing the third letter in the DNA word. To determine the number of 3-letter words you have to count the total number of lines connecting the second with the third letters.

BOX 5 Codons

Three letter words play a very important role in deciphering the DNA code. You learned earlier that your body uses the information in DNA to build proteins. The building blocks of proteins are the **amino acids**. Groups of three adjacent nucleotides (3-letter words called **codons**) code for an amino acid, the building block of proteins. There are 64 possible codons in DNA, but there are only 20 amino acids. Different codons must therefore code for the same amino acid. (Workers in this field—who are often very fond of fancy names – call this the *degeneracy of the code*)[a].

---

[a]We have listed the names of all codons and the amino acide for which they code in Table C.1 in appendix C

- Each group counts the number of words which they found.

- Add your numbers.

- What is the total number of words you found?

- Is there a way to predict the total number of possible words?

- Did you find all possible words? Did you miss one or more? If yes, which ones did you miss?

- Compare your solution with the solution in appendix A. ◇

Figure 4.4 shows a graphic representation which may help to identify all possible 2- and 3- letter words without missing any of them.

Three letter words play an important role in DNA sequences (see also Box 5).

Figure 4.4:

a
a
a
c
g
t
c
a
c
g
t
g
a
c
g
t
t
a
c
g
t
c
a
a
c
g
t
c
a
c
g
t
g
a
c
g
t
t
a
c
g
t
g
a
a
c
g
t
c
a
c
g
t
g
a
c
g
t
t
a
c
g
t
t
a
a
c
g
t
c
a
c
g
t
g
a
c
g
t
t
a
c
g
t

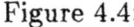

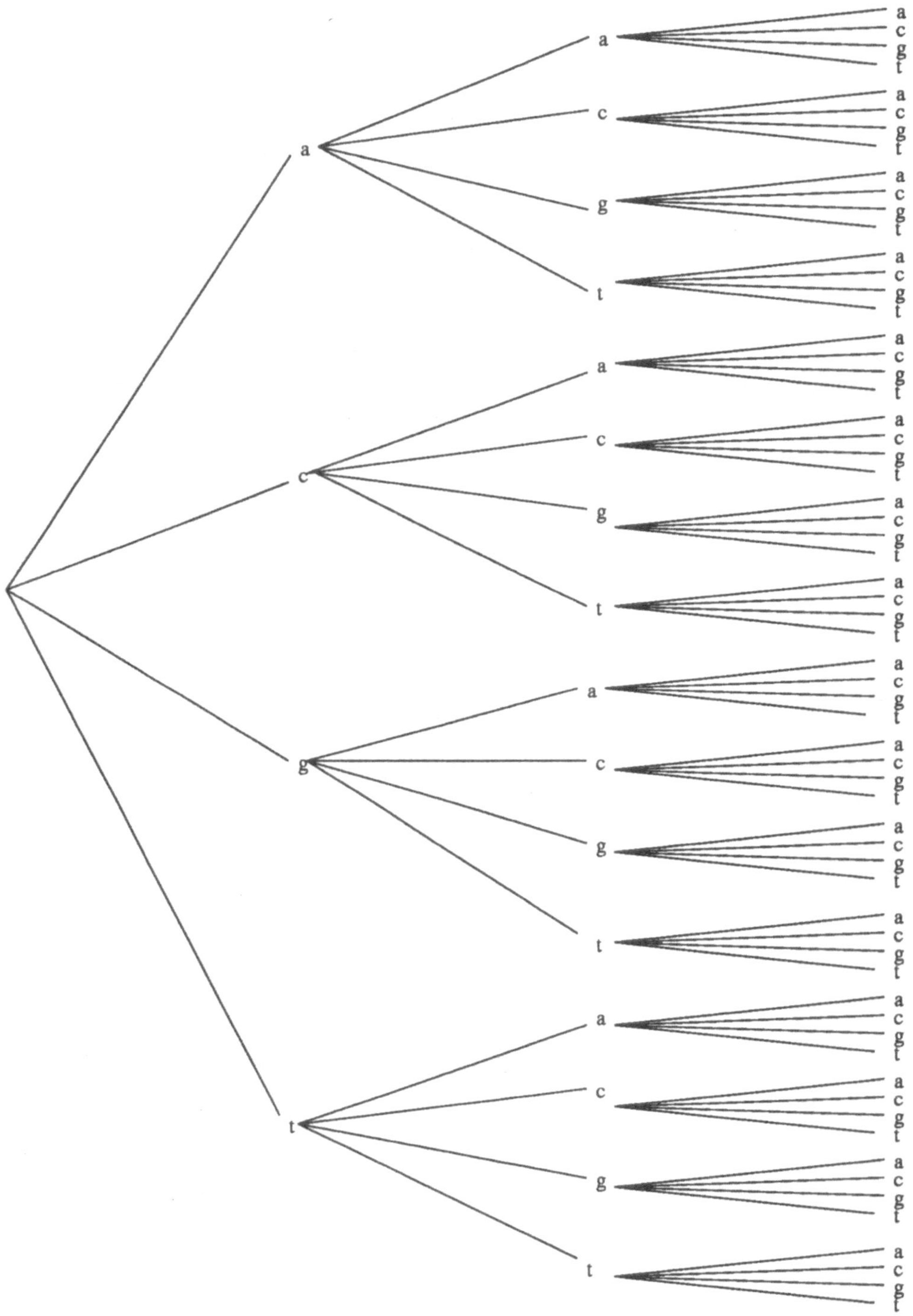

◇ **Activity 10:   4-letter words**

How many different 4-letter words can be constructed from the letters A, C, G, and T? To find the answer you have two options:

1. Sit down and spend the whole afternoon writing down all the possible 4-letter words and count them. This is a tedious job.

2. Derive a general rule from the entries in Table 4.8. Maybe the tree structure in Fig. 4.4 can help you find a general rule. What would you have to do to expand the tree structure to 4- or 5- letter words?◇

Table 4.8: Possible number of n-letter words. Activity

Word length = number of letter	Number of different words
1	4
2	16
3	64
4	
5	
⋮	⋮
n	

4.3.5   Zipf's Law in DNA

Before we start analyzing real DNA sequences we will first try a random sequence of nucleotides:

◇ **Activity 11:   Words in a Random Sequence**

- Generate a random sequence of the four letters a,c,g, and t by using the **Random DNA** option of the program.

- Choose a word length n (n = 3, 4, 5 or 6) and count the number of occurrences of n-letter words in the random sequence.

- Plot your results in a Zipf plot and determine the slope.

- Write your results down in Table 4.9.

- Generate a second random sequence of DNA letters and analyze this sequence. Enter the result in the table.

- Generate more random sequences and add your results to the table

- Compare your results with results of your friends. Add their results to your table.

Table 4.9: Word Occurrences in Random Sequences—slopes of Zipf plots. Activity

sample	slope
1	
2	
3	
4	
5	

- Can you find an explanation for the slope? Discuss with your friends!

- Do not turn the page before you have answered the question!◇

Did you find a slope close or equal to ~ 0? The explanation is the following: In a random sequence, all letters occur with the same probability and the probability of finding a letter at a certain position does not depend on the preceding letters. Therefore the number of occurrences for a certain n-letter word in the sequence is almost the same for all words. It would be exactly the same for an infinite sequence.

◇ Activity 12: Word Occurrences in DNA

- Try the same with real DNA sequences. As a start we suggest the following sequences

    1. LAMCG[13] and/or PODOT7[13]
    2. HUMRETBLAS[13] and/or HUMNEUROF[13]

- Determine the number of word occurrences for each sequence.

- Make a rank plot and determine the slope. Write your result in Table 4.10

- Split the sequence up into coding and non-coding parts

- Determine the number of word occurrences for the stitched-together coding and non-coding parts separately.

- Make rank plots and determine the slopes for coding and non-coding parts.

- Write your results into Table 4.10.

- Try other sequences from[14]

    - the B.U. diskette
    - CD-ROM
    - Internet (from the group at B.U., WAIS Server)

- Write your results for these into the table.

- Compare them with the results of your friends.

- Can you find a trend for the slopes of the whole sequences? For the coding parts? For the non-coding parts?◇

Do not turn the page until you have finished activity 12.

---

[13]These are the names under which the DNA sequence is saved on the database.

[14]Your teacher will know which of these materials and connections are really available

Table 4.10: Slopes of Zipf plots for DNA. Activity.

DNA-sequence	slopes		
	whole sequence	coding	noncoding

Were the slopes for the DNA sequences clearly larger than for the random sequences? That shows that a DNA sequence is in general different from a random sequence. Could you verify that the slope for the coding parts is always smaller than the slope for the non-coding sequences? This is an indication that the coding parts behave more like a random sequence than the non-coding parts. This finding will become very important in our next chapter.

4.4 Landscapes of Literature and DNA

4.4.1 Why a new approach?

It is useful to approach a scientific problem from more than one direction and with diverse tools. Workers in different locations often work on the same problem using different methods. Each method typically results in different kinds of data and leads to insights that complement those resulting from other methods.

Each worker tends to use techniques with which he or she is already well acquainted. In the preceding chapter you learned about Zipf's Law to characterize languages and DNA. The following method to describe the features of languages—natural languages or the DNA language of your body—was developed by workers who worked on random walks and "landscapes". Guess what they did, when they started to work on languages! They constructed "random" walks and landscapes from written texts and DNA sequences.

Therefore, let's *Climb the "language mountain"!*

4.4.2 Climb the "language mountain"

You may wonder how a random walk can be constructed from a written text.

Recall the random walk unit. Fig. 4.5 summarizes the different steps which leads to the construction of a walk profile. Construct the random walk of an ant along a line by flipping a penny: for each head the ant steps right and for each tail it steps to the left(Fig. 4.5.A). The position x of the ant is increased or decreased in each step(Fig. 4.5.B) In order to construct the profile which belongs to the random walk, you plot the position of the ant after each step against the step number and connect the points by straight lines (Fig. 4.5.C)[15]. Fig. 4.5.D shows the resulting profile after 5 steps. In Fig. 4.6 you can see part of a larger landscape.

Let's now imagine that you want to compare your ant's walk with your friend's. Somehow you have to save your results. One way would be to write down the sequence of heads and tails when you flipped the coin, e.g.

> H H H T T H T T T.....(and so on)

---

[15] By switching the step directions you can create an exact mirror image of the profile.

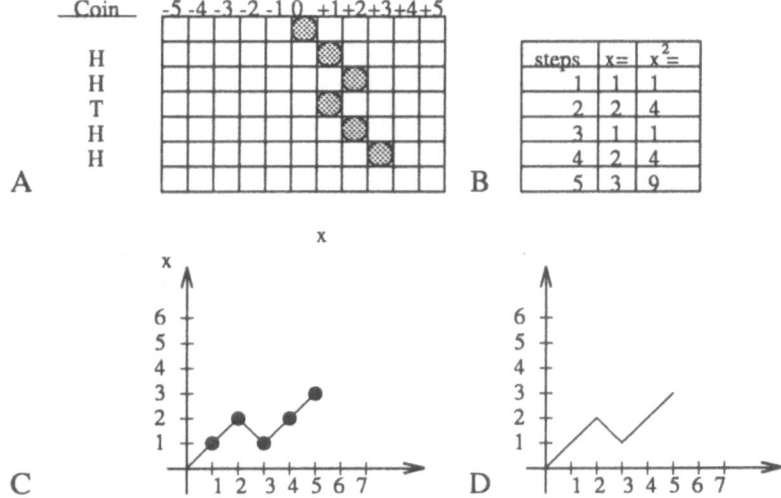

Figure 4.5: Constructing a profile (or landscape) from a . A: The ant steps to the right or to the left depending on the outcome of the coin flipping. B: Listing the displacement of the ant after each step in a table. Also calculated is the square displacement. C: Plotting the displacement of the ant vs. the step number and connecting the points with straight lines. D: The resulting walk profile.

Random Walk Profile

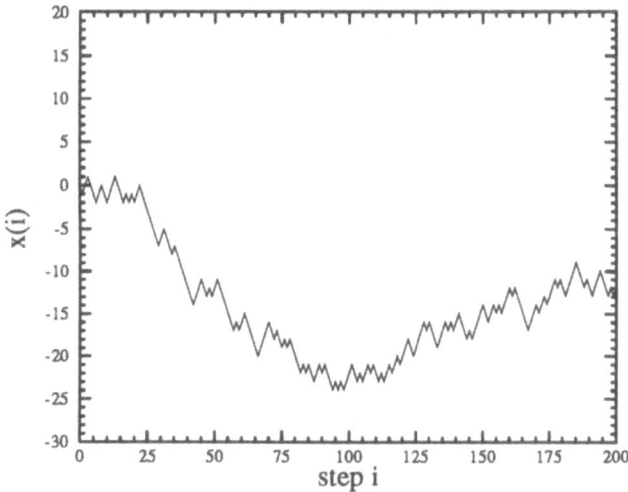

Figure 4.6: Example of a Random Walk Profile.

This "text" contains all the information which you need to reconstruct your random walk or its profile. From another point of view you can say, that the "text" in the box is represented by a random walk profile. Different sequences of heads and tails (different "texts") lead to different random walk profiles. Therefore, one way to compare different sequences of letters is to construct and compare their walk profiles (or landscapes). The walk profile becomes the **"fingerprint"** of the text.

Of course the letter sequence above presents a very poor example for a text: It is built from a two-letter alphabet and contains no spaces, periods, etc. The question is: Does there exist for real texts and languages a fingerprint landscape such as the one described above? Wouldn't it be great to find a tool which enables us to distinguish between, e.g., different authors just by looking at some "text profile", i.e., a graphic representation? Or a tool which helps us to characterize coding and non-coding parts in DNA sequences?

4.4.3   DNA landscapes

You will see now that it is not very difficult to construct a "fingerprint landscape" for DNA sequences because we can do for a nucleotide sequence almost the same thing that we did for the text made up of heads and tails. Not a sequence of heads and tails in a coin flipping experiment, but the sequence of bases in a nucleotide sequence guides our ant on its walk in one dimension. Consider a nucleotide sequence of the four bases adenine, cytosine, guanine, and thymine, e.g.

accgtttaa gcctgattac ...

From the beginning of the sequence to its end, each base tells our ant how to step. But there is a problem:

In our coin flipping experiment the outcome could be either head or tail. Depending on the result the ant stepped either to the right or to the left. Now we have four different bases in the DNA sequence, but our ant can only understand "step right" or "step left". Therefore, in order not to confuse our ant we have to "filter" or to "translate" the nucleotide sequence into "rights" and "lefts". We do this by dividing the nucleotides into two groups. The ant has to step right whenever the nucleotide is in the first group or to step left whenever the nucleotide is in the second group. We can for instance choose that the ant steps right for each A and left for all other bases or that it steps to the right for a pyrimidine base, and to the left for each purine base. We call this set of instructions **walk rules** or **step rules**.

Can you find other so called "step rules"? How many different step rules can you find?

◇ **Activity 13:   DNA Walk Rules**

- Continue filling the Table 4.11 ◇

	step left for base	step right for base
1	A	C, G, T
2	C	G, T, A
3		
4		
5	A, C	G, T
6	C, G	T, A
7		
8		
9		
10		
11		
12		
13		
14		

Table 4.11: DNA Walk rules. Activity

How many different rules did you find? By applying one of these rules to our chromosome sequence we can create a landscape for our sequence when we plot the successive positions of our ant. Can you see that for every rule there is another rule which produces an exact mirror image profile? You can find all possible walk rules in app A.

Our goal remains to construct a "fingerprint profile" for DNA sequences which helps us to distinguish between coding and non-coding regions. By constructing landscapes for many different DNA sequences from all of these rules workers found that especially for one of the rules—the so called **purine vs pyrimidine rule** which means "step down" for purine (a and g), "step up" for pyrimidine (c and t)—the landscapes for coding and non-coding regions sequences show clear differences. Therefore in the following activities we will use the purine-pyrimidine rule exclusively.

◇ **Activity 14: DNA landscapes**

We will now use the computer to produce landscapes of DNA sequences.

- Load the DNA sequence called LAMCG[16] into your program[17].

- Construct the profile by clicking on DNA walk[18].

- Do the same with the HUMNEUROF-sequence.

- Compare the two profiles! Can you find any differences? Can you observe properties they have in common?

- Create or load a random walk sequence.

- Construct the profile.

- Which profile of the LAMCG- and HUMNEUROF-sequences most closely resembles the profile of the random walk?◇

The profile for the HUMNEUROF-sequence has a more jagged contour than the profiles for the LAMCG-sequence and the random sequence. The difference between the two profiles is a very important one, because it provides us with a tool to distinguish between sequences: The LAMCG-sequence is a sequence which mainly consists of coding regions. It is 87% coding. You can check this with the option "fraction" in the DNA menu. The sequence HUMNEUROF is mainly (88.6 %) non coding.

If you use the rule, the profiles for chromosome sequences which contain a huge amount of "junk" will always appear to be rougher than the profiles of coding sequences. Therefore we can use the purine - pyrimidine rule to construct a "fingerprint profile" for DNA sequences.

Do you want to learn how to express the roughness of the curves mathematically? If not, skip the next section 4.4.4 and continue on page 209 of section 4.5.1.

4.4.4 The roughness exponent

There is even a way to express mathematically the "roughness" of the profiles.

Figure 4.7 shows two profiles like the ones you can construct from DNA or written texts. In order to find a mathematical expression for the roughness of the profiles, we first have to describe what we mean, when we say, the profile is rough.

---

[16] *See* footnote 13 on page 185.

[17] Your teacher will know if you can get it from the B.U. diskette, from CD-Rom, via ftp from the B.U. group, or from the wais server.

[18] In the random walk unit you learned that it is often necessary to subtract the overall bias in order to see something else than just a straight line. Our program will subtract the bias for you

◇ Activity 15: Two Profiles

- Try to describe "roughness". Tip: compare Figs. 4.7.a and 4.7.b on the following page.◇

The profile in the Fig. 4.7.a appears to be smoother because the "mountains" in the profile do not rise as fast or slope down as rapidly within a certain length scale as in Fig. b. Unfortunately, this visual criterion is not very precise and the human eye can easily be deceived. For instance, if we plotted Fig. 4.7.a on a smaller y-scale, the profile would seem to be much rougher. Therefore we need a more "objective" criterion to characterize the landscapes. In the following we will derive a number which can help us describe the landscapes. In order to derive this quantity, we look at the profile in figure 4.8 which is identically with the profile in Fig. 4.7.b.

When we want to describe a part of our profile—for instance the part in the little box on the left in Fig. 4.8—we can mention the height $W(l)$ which the box of length l must have in order to cover the little part of the profile. $W(l)$ is the distance between the minimum and the maximum of the profile in the box (cf. Fig. 4.8). characterize our profile, because the steeper the profile in the box, the larger is $W(l)$ for a constant value of l.

However, if you look at another part of the profile, you will probably find $W(l)$ to be different from $W(l)$ in the first box. In order to describe our profile we have to measure $W(l)$ for many different locations along the profile. Thus, for a very long profile it would be a very tedious job to write down the difference of height for each little box. Fortunately, we can define a reasonable quantity $\overline{W(l)}$ by averaging the height over all boxes.

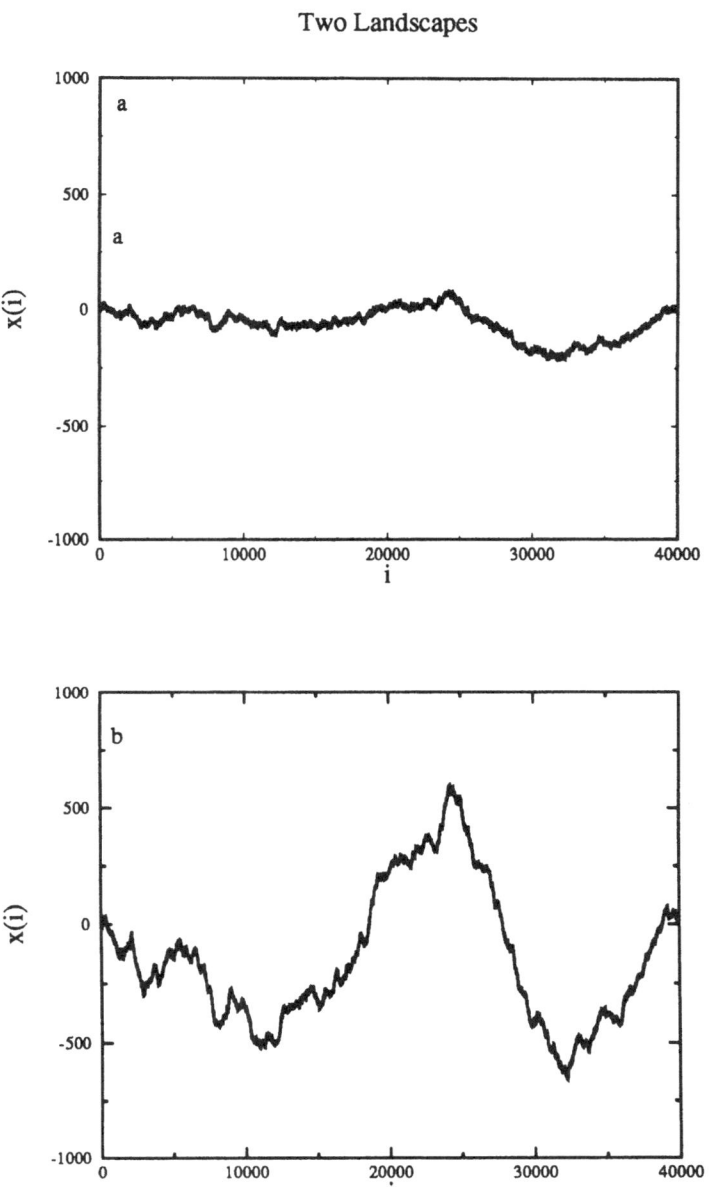

Figure 4.7: Two different profiles. Can you describe the differences?

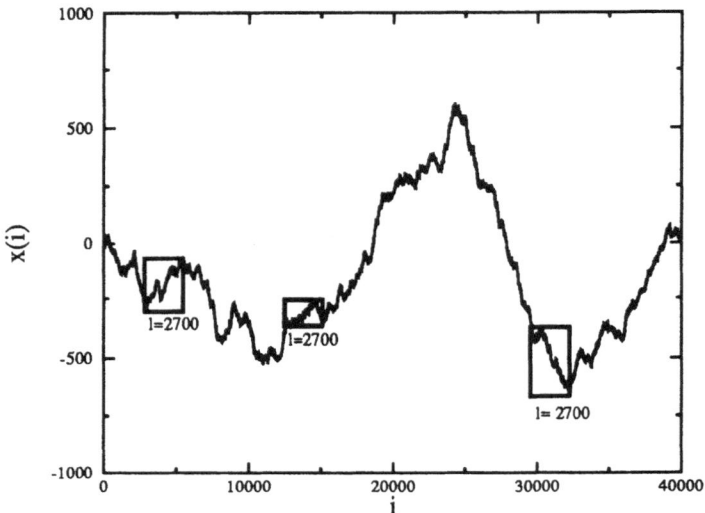

Figure 4.8: The profile is the same as in Fig. 4.7.b. Three boxes for the measurement of $W(l)$ at different parts of the profile are drawn. The length of each box is 2700 nucleotide positions. Though the boxes have the same length, their heights $W(l)$ are different for all three boxes.

◇ **Activity 16:   Measuring $W(l = 1)$ for three profiles.**
On the next pages (196, 197, 198) you will find three different profiles.

- Draw your profile in Fig. 4.11

- Cover parts of the profiles (Fig. 4.9 –4.11) with boxes of length $l = 1$ and measure $W(l = 1)$ for each box.

- Determine the average value of $W(l = 1)$ for each of the profiles[19].◇

---

[19] Please, read the text in Box 6

BOX 6 Average Values

Do you remember from the random walk unit that workers like to average over the square of a quantity? In this box we will give you a short example: Assume that we measured for a profile the following values for $W(l = 1)$: 1, -1, -1, 1, 1, 1, -1, 1, -1, -1.
We then calculate $\overline{W(l = 1)}$ as:

$$\overline{W(l = 1)} = \sqrt{\frac{1}{10}[1^2 + (-1)^2 + (-1)^2 + 1^2 + 1^2 + 1^2 + (-1)^2 + 1^2 + (-1)^2 + (-1)^2]}$$
$$= 1$$

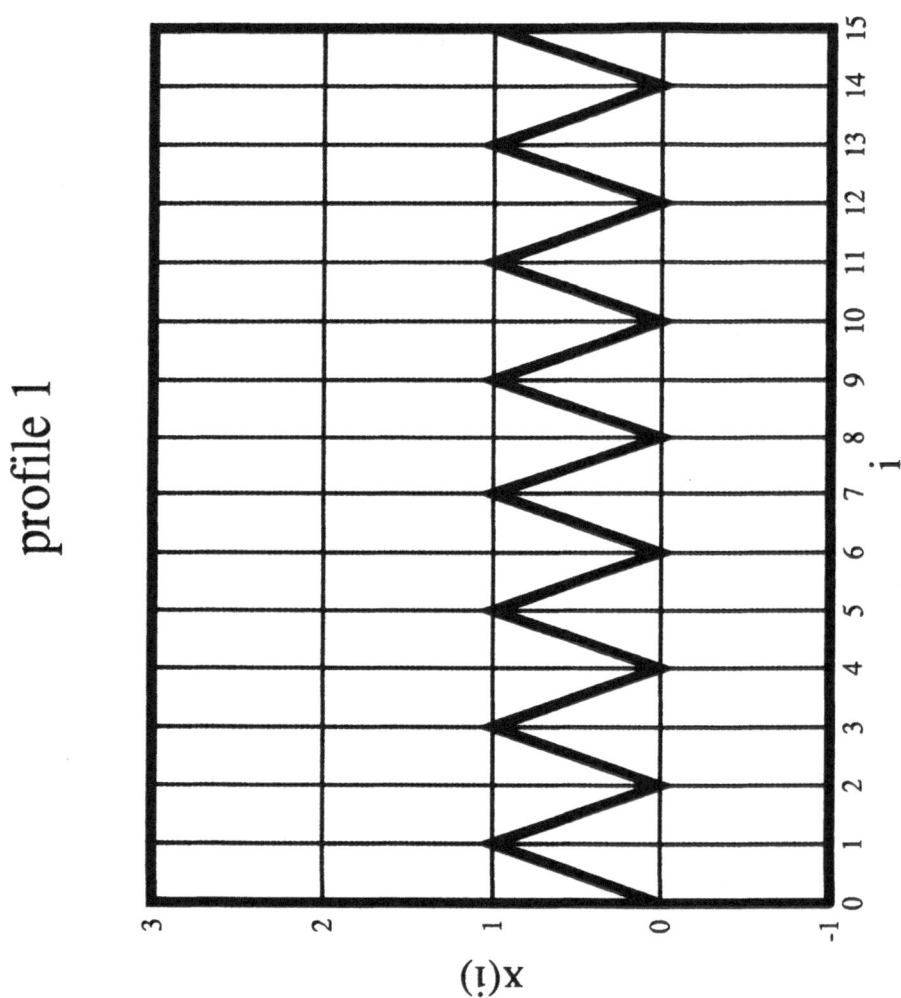

Figure 4.9: Profile i.

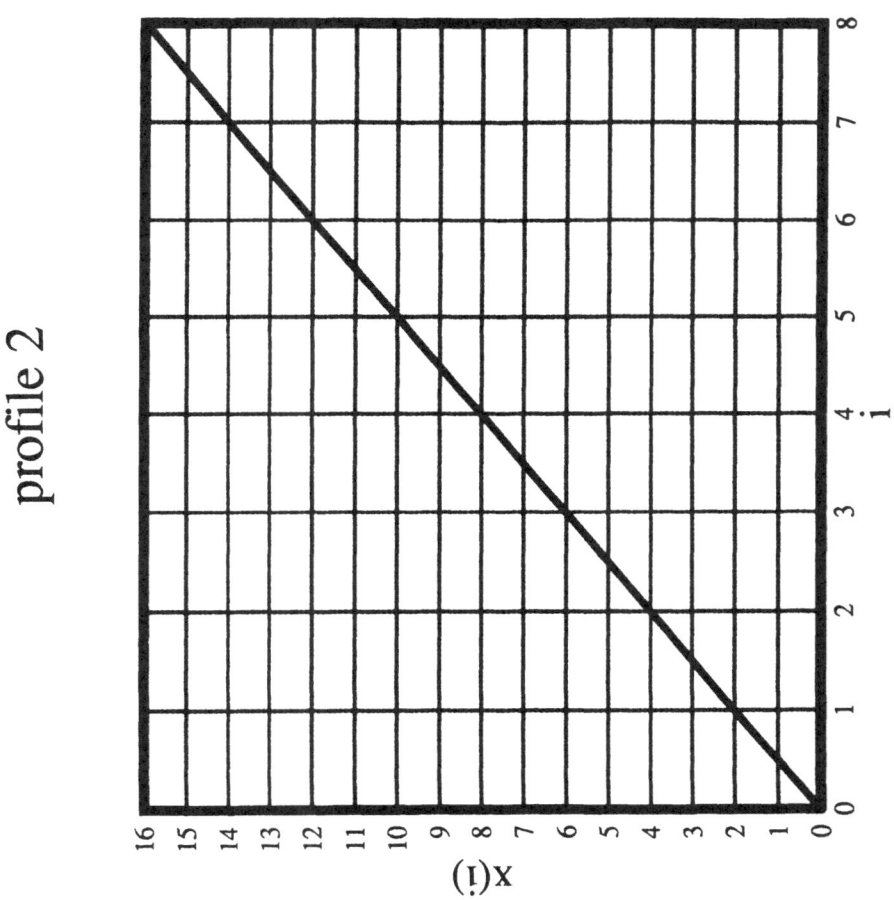

Figure 4.10: Profile ii

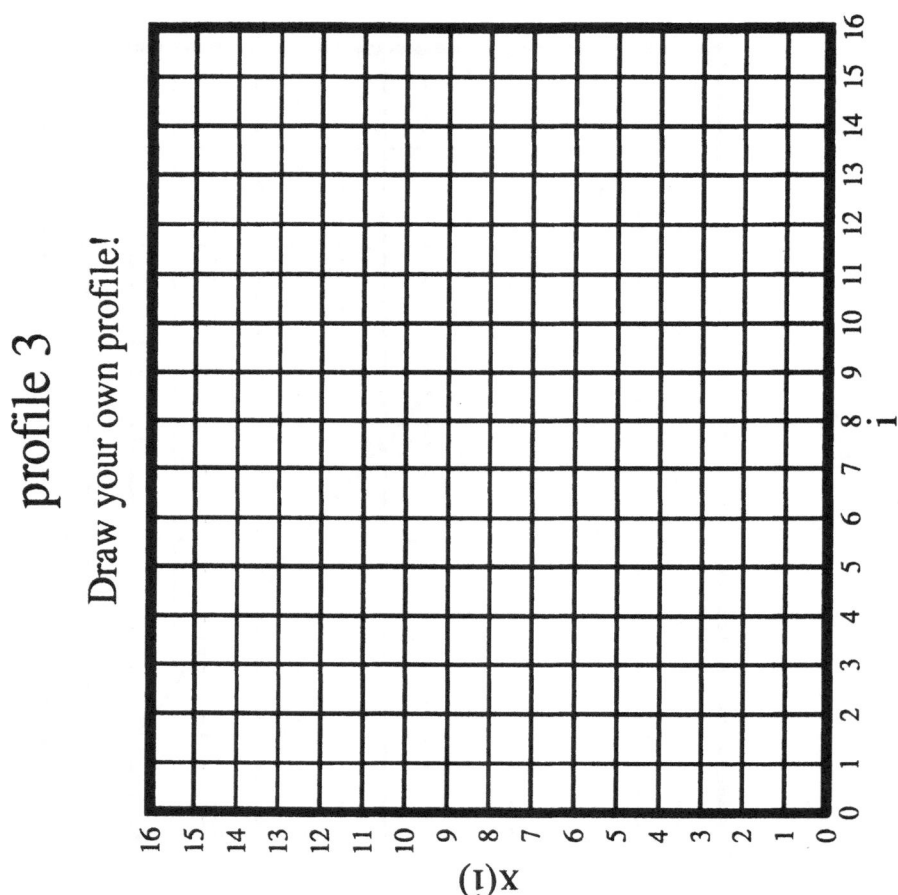

Figure 4.11: Profile iii: Draw your own profile! But remember: In each step you can only step up or down by 1!

You have measured $\overline{W(l=1)}$ for different profiles. However, we may ask what $\overline{W(l)}$ really tells us about the profile: For the profiles Fig. 4.9 and Fig. 4.10 $\overline{W(l=1)}$ is equal 1 for both profiles though the profiles are really different. But we still think that $\overline{W(l)}$ is a reasonable quantity. The secret lies in varying the size of l.

◇ **Activity 17:   Measuring $\overline{W(l)}$ for different values of *l***

- Repeat the measurements in 16 for $l = 2, 3, 4, ...$

- Write your results for $\overline{W(l)}$ in Table 4.12.◇

When you now compare the values in Table 4.12 you see that $\overline{W(l)}$ varies differently with *l* for different profiles.

By measuring $\overline{W(l)}$ for different profiles and different values of *l*, workers have found that for long profiles $\overline{W(l)}$ follows the rule:

$$\overline{W(l)} \propto l^{\alpha} \text{where } \alpha \text{ is a real number between 0 and 1.} \qquad (4.1)$$

When you have worked with programs like the *Fractal Coastline* or *Fractal Dimension*, expressions like eq. 4.1 are familiar to you. Maybe you remember that data points that can be described by such a formula are lying on a straight line when plotted in a log-log-plot. The slope of the straight line is in our case the exponent α in eq. 4.1.

◇ **Activity 18: Exponents for Profiles i, ii, and iii**
We will try now to find the exponent a for the profiles i) to iii).

- Plot your data from Table 4.12 on log-log paper[20].

- Calculate the slope for each graph. Write your results into the table and compare with the solution in app. A.

Profile	Exponent

- Don't turn the page before you are done with this activity.◇

---

[20] If you feel more comfortable with using logarithm, plot log $\overline{W(l)}$ vs log *l*.

Table 4.12: Measuring of $W(l)$ for different profiles. Activity

Profile 1

l	$\overline{W(l)}$

Profile 2

l	$\overline{W(l)}$

Profile 3

l	$\overline{W(l)}$

We find that for profile i, $W(l)$ is a constant and independent of l. Therefore the exponent is 0. For profile ii we find that $W(l)$ is proportional to l which means that the exponent is 1. Profile i) and profile ii) represent special cases of profiles where α becomes exactly one of the limiting values 0 and 1.

What did you find for your profile iii?

The exponent α which we call the **roughness exponent** of the profile is in fact a very important quantity. It helps us characterizing the profiles. However, workers sometimes prefer a slightly modified definition of the scaling exponent[21]. Before they measure the height $W(l)$ of a box, they subtract the local bias of the landscape in the box[22]. Thus they make sure that only really rough profiles get high scaling exponents(between 0.5 and 1).

We really have found a characteristic to distinguish between profiles: The rougher the profile, the higher is the exponent. For instance the two profiles in Fig. 4.7 have exponents 0.5 (a) and 0.75 (b). Profiles that have roughness exponents close to 0.5 are not entirely unknown to you: These are profiles of random sequences.

◇ Activity 19:   Roughness Exponents for Random Sequences

We will now look at different kinds of sequences to see how we can characterize them by measuring the scaling exponent α.

- Let the computer generate a random nucleotide sequence.

- Construct the landscape and calculate the scaling exponent.

- Write your result into the table.

- Repeat the procedure at least two more times.

- Compare your results with the results of your friends.◇

Sequence	exponent	exponent	comments
Random sequence 1			
Random sequence 2			
Random sequence 3			

---

[21]...just another name for the roughness exponent.

[22]They do for a little part of the landscape exactly what you learned to do for a random walk profile.

For the random sequences you should have found values approximately equal to 0.5 for the exponent.

◇ **Activity 20: Scaling Exponents for DNA sequences**

- Calculate now the scaling exponent α for all the DNA sequences for which you have constructed the random walk profiles earlier.

- determine the fraction of coding and non-coding material in the sequence

- Write your results into the table:

sequence	exponent	% coding parts	% non-coding parts

- Compare the exponents for mainly coding and mainly non-coding sequences. Do you find differences?

- Do not continue unless you are done with the activity.◇

You found exponents close to 0.5 for sequences that contain a large number of coding regions. For sequences that are mainly non-coding the exponent is clearly higher than 0.5. An exponent of 0.5 for coding sequences implies that coding sequence in DNA behave like random sequences because there we got also an exponent of about 0.5. his is the second indication[23] for the *random character* of coding regions in nucleotide sequences. The higher exponent in non-coding parts tells us that the non-coding parts of a nucleotide sequence do not at all behave like a random sequence. These parts are behaving like so called *correlated sequences*. The difference between random and correlated sequences can be explained as follows: When you construct a random sequence, for instance by flipping a coin for 100 times, the outcome of the 100th

---
[23]cf page 187

flipping[24] does not depend on the 99 earlier results. However, in a correlated sequence the outcome *does* depend on preceeding results.

◇ **Activity 21: Goody!***
The random sequences are constructed in our computer by a random number generator. This is an algorithm which returns numbers (real or integer, depending on the algorithm) in an order, as random as possible. To write a good random generator is a very difficult task. We will now try, if a "human being is a good random number generator".

- Type your own random DNA sequence: Use only letters A, C, G and T.

- Construct a sequence of at least 1000 nucleotides. You can work together with friends. Each of you can construct a portion of the sequence. Now construct the profile and calculate the exponent α.

- Compare your results to the results of other groups! Is α equal to 0.5 , or clearly smaller or larger than this value? Why?

Often when people are asked to write down "random sequences", they think that in order to write down a random sequence they have to avoid repeating symbols. The results are sequences which are not really random. They have an exponent α that is lower than the value 0.5. Workers call these sequences *anti-correlated.*◇

4.4.5 Zero Information or Maximum Information?

We now have what seems to be a paradox. A totally random sequence of four letters A, C, G, T, generated using a random walk and carrying no information whatsoever, has a scaling exponent $\alpha = 0.5$. The coding regions of DNA, carrying all the information needed to create human proteins, also have a scaling exponent $\alpha = 0.5$, and act as if they too were generated by a random walk.

What is going on? The answer is profound and goes to the heart of both DNA structure and information theory. Start with the following story:

---

[24] head or tail in a coin flipping experiment

Assembling a Bicycle

Your are assembling a bicycle from its individual parts. You have a set of step-by-step instructions that tells you what to do next in complete detail. However, you have same experience with bicycles. You know, for example, what an assembled bicycle looks like. So the instruction "Put the rubber tire on the rim of the wheel" is not really necessary for you; you can guess with fair certainty when it is time to carry out this step. Many other instructions in the assembly process you also do not need.

Now try another set of instructions. A close friend, who is a professional bicycle assembler and knows you extremely well, makes out a set of instructions specifically for you. She leaves out all instructions that you know or can guess, giving you only those instructions which you would have no chance of predicting. Then before reading her next instruction, you have *no idea* what that instruction will be. From your point of view, the next instruction is completely random; there is no way for you to predict it.

A very similar situation applies to the coding sequences of DNA. The most efficient possible code is one whose next letter is completely unpredictable. If you are an enzyme using the DNA code to assemble a protein molecule, *the maximum possible information is carried by the next letter when you cannot predict that next letter.* If you know or can guess the next letter, you do not need the next letter, so its information is wasted (or partially wasted).

In brief a DNA sequence that is the most efficient possible carrier of information is indistinguishable from a completely random sequence, as far as the scaling exponent α is concerned. Both sequences have the profile of a random walk and the scaling exponent $\alpha = 0.5$.

So of what use is the scaling exponent analysis of the coding regions of DNA? Of no use if we already know they are coding regions. But the scaling exponent may help us *find* the coding sequences.(*See* Section 4.5).

And the non-coding sequences? Their scaling exponent is different form that of a random walk. What does this mean? We do not know. That is the excitement of current research. One possibility is that the non-coding sequences are summaries or dictionaries of instructions rather than the instructions themselves. Possibly they organize or supervise the replication process in some way. And here is another little story.

The Telephone Book

Think of a telephone book for a large town or small city. In this town every telephone number begins with either 123 or 456, each followed by four more digits. In telephone parlance, the town has two telephone *exchanges*. When you look up a number in the telephone book, you know a lot about what comes after each name: either the digits 123 or the digits 456. After you read the first of these digits you know even more: exactly what the following two digits will be. So the telephone book is neither a random set of symbols nor is it the most efficient code for presenting telephone numbers.

How could you make the telephone book for this city more efficient in coding information, thus reducing the size of the telephone book and saving money for the telephone company? Would your "more efficient" telephone book be easier or harder to use?

The non-coding sequences of DNA may be like this telephone book: inefficient listings or summaries of same kind. That is what some people are trying to find out.

4.5   Discover coding regions in DNA

The Beachcomber Algorithm

You should have found that DNA sequences which are mainly coding have a smaller roughness exponent α than non-coding sequences. In the following we will try to use this finding to locate possible coding regions. How can we do that? How can we find in a DNA sequence, like the one you can see in Table 4.1, those parts which differ from the rest of the sequence because they code for proteins?

Assume for a moment that we know which parts in the sequence of Table 4.1 are the coding parts. Then we can construct landscapes for each coding and for each non-coding subsequence. When we compare the "finger print landscapes" of the coding parts with the landscapes of the non-coding parts, we find these landscapes to be different from each other. The exponent α calculated for the coding parts (0.5) would be smaller than α for the noncoding parts (0.6).

Thus, you can see that we cheated a little bit before. When we calculated α for a whole DNA sequence, we pretended α to be uniform over the whole sequence. But this is not always true, because as we learned before, a DNA sequence usually contains coding *and* noncoding parts; so different parts of a DNA sequence can have different values of α.

Coding parts have lower values for α than noncoding parts. We can use this difference in α to help distinguish between coding and noncoding sequences of DNA.

We can solve the "dark secrets of a DNA sequence" and "unmask" its coding parts: Like a detective who looks at fingerprints to unmask the criminal, we can look at a DNA sequence to find the coding sequences!

Our program does the following: It takes a DNA sequence and constructs the profile. Then it takes the profile up to nucleotide position 1000 [25], and calculates α for this part of the sequence. Fig. reffig:firstpoint illustrates the algorithm.

One way to proceed is to move the box along the sequence to the next 1000 nucleotides and calculate alpha again, repeating this procedure until we reach the end of the sequence. After having covered the whole sequence with little boxes and having calculated alpha, we could gain a table like Table 4.13 of values for alpha for the whole sequence[26]:

Now we can compare values in the table. For all values of α close to 0.5, we can assume that the subsequence is part of a coding region in the DNA

---

[25]1000 is a good number because many coding regions are of the size of 1000 or more nucleotides

[26]These are really computed values for a DNA sequence.

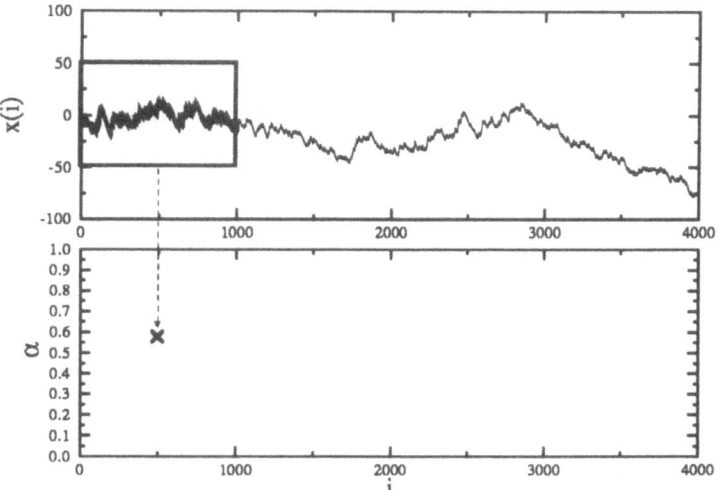

Figure 4.12: Calculating the first point for the beachcomber plot. The box in the picture has the length 1000 nucleotides. The part of the sequence analyzed is the thick line in the box, for which the program calculates alpha. In this case $\alpha = 0.57$.

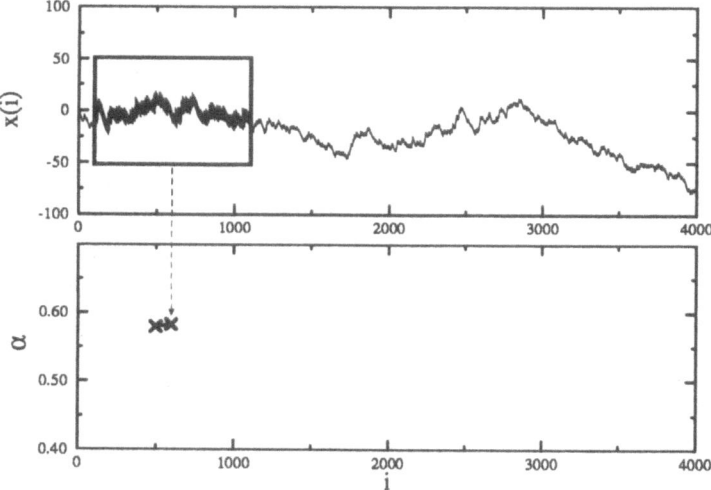

Figure 4.13: Calculating the second point for the beachcomber plot.

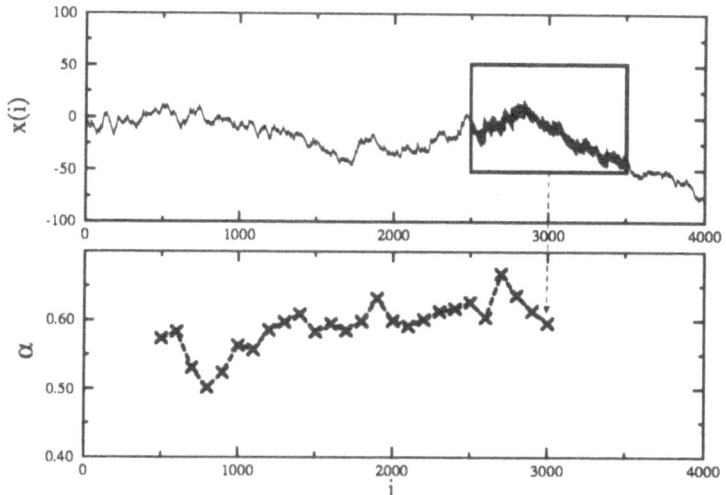

Figure 4.14: Calculating more values for α for the beachcomber plot.

nucleotide position	α
1-1000	0.57
1001-2000	0.58
2001-3000	0.63
⋮	⋮
5001-6000	0.51
⋮	⋮

Table 4.13: Values of α in a beachcomber plot.

sequence. And indeed, the nucleotides 5001 to 6000 are part of a large coding region in this sequence.

But what would happen if a coding region is "hidden" between the 600th and the 1600th nucleotide, for example? We could not find it, if we moved the box by 1000 nucleotides. One way around this is to move the box by a smaller distance and calculate α again. One possible choice is to move the box by just one nucleotide. Unfortunately, DNA sequences are usually very long and our program would need an awfully long time to calculate all these values. Moreover, a sequence in the box differs from its predecessor sequence only by 2 nucleotides (one lost at the beginning, one gained at the end). The α for the new sequence is then about the same as the α for the first sequence. We would not gain much new information.

Therefore, we try the following compromise: We move the box by a distance of, say, 100 nucleotides and calculate α again. But what are we going to do with all these values for alpha? Write them in a table? As you might have noticed, a graphic is easier to analyze, because a good graph can make details of the data more visible. We will plot α as a function of the nucleotide position. But where do we put the α we have calculated for the first 1000 nucleotides? We plot it at the middle of the box, at nucleotide position 500(Fig. 4.12). Then we move the box one hundred nucleotides, calculate α again and plot it at the center position of the new box, at position 600 (Fig. 4.13). By sliding the box along the whole sequence we can get a curve like Fig. 4.14.

The curve in Fig. 4.14 drops to the value 0.51 at approximately position 800 for the center of the box. This is evidence for a coding regions in the neighborhood of this position. For many DNA sequences the position of dips in the curve and the positions of coding regions are in very good agreement.

This way of detecting coding regions is called the **beachcomber method**, because a beachcomer walks along the beach looking ahead and behind him/her for interesting articles washed up by the tide, thus sampling the contents of a "box" with him/her at the center.

$\diamond$ **Activity 22: Try to find the coding sequences in different DNA sequences:**

- Calculate the α curve for different DNA sequences using the beachcomber menu of program **Landscapes**.

- Find the "dips" of the α curve.

- Choose the "show coding option" in the beachcomber menu.

- Compare position of the dips with the coding regions.[27]$\diamond$

---

[27] Coding and non-coding region will be displayed in different colors.

Discussion

You may have found DNA sequences for which the algorithm works very nicely, whereas for others there was no correspondence in the positions of the dips and the positions of the coding regions. If there was no correspondence, the problem was either that there was a coding region without a dip in the curve or a dip appeared in a non coding region. You find both failures e.g. for DNA from humans and other high organisms. These are the two main problems: a) The coding regions in human DNA sequences are very short. We cannot find them by using a box of 1000 nucleotides in our algorithm. But it is not possible to choose much smaller boxes, because than the box is too small to calculate a reasonable α because of statistical fluctuations. b) You can find sequences in the non coding regions of a human DNA sequences which might have been coding at some time in the evolutionary past. Today these are not coding regions, but they still possess properties of coding materials, e.g. a value of α about 0.5.

4.5.1 Landscapes in Literature

In the last section we found out that landscapes are a good tool for learning about DNA sequences. It would be nice if we could have a similar success in applying the same method to literature. The major difference is now, that we have a 26-letter alphabet or a 32–letter alphabet if we include spaces and punctuation. These numbers can vary from language to language because of the special characters or accents which are used in other languages than English: for instance "ñ" in Spanish, the umlauts "ä, ö, ü" in the German language, letters like "æ" in Danish, etc. In Portuguese the letters "k, y, w" which often appear in English words are not used at all. In other parts of the world (Eastern Europe, Asia, Arabia) different types of alphabets are used. But this is essentially no problem for constructing landscapes in literature because we can adjust the walk rules - which we have yet to define - to the language in which a text is written.

Because we have now many more "letters" in our alphabet than there are in the DNA alphabet, we also have many more possible walk rules[28]: walk rules that tell us for which letters we have to step up or to step down on our literature walk.

$\diamond$ **Activity 23: Number of Walk Rules for a 26-Letter Alphabet**

- Estimate the number of possible walk rules.$\diamond$

---

[28] Remember: For the DNA walk we found 14 (7, if we exclude mirror images) different walk rules.

You find the exact number in app. A.

It is clear that for such a huge number of walk rules no one can use every one to analyze a given text. Actually, as far as we know nobody has ever used this kind of walk rule to study properties of literature. Therefore, you are now at the front edge of research. You may find results which nobody has ever discovered before you.

◇ Activity 24: Literature Walks

The computer program **Landscapes** is equipped with tools to read a text, to define a walk rule, to construct the landscape and to calculate a scaling exponent for the landscape.

- Use this tools do develop your own research program, working in groups:

- Before you start analyzing landscapes in literature, write down your questions, ideas, expectations about the literature walks on the next pages[29].

    1. Write down step by step—as for previous activities—what you want to do with the program. Which walk rule do you want to use? Which texts do you want to analyze?

    2. Write down what outcome you expect.

- Carry out the steps which you have defined in step 1. Write down your results, questions, comments which arise during the activity.

- Compare your results with your expectations (step 2).

- Summarize your findings.

- Hold a "research conference" in which groups report their results and everyone joins in discussing the meaning of these results.◇

---

[29]In appendix A we have listed some possible the questions.

Appendix A

Answers to problems.

Activity 1: AT 100 crosspieces per page, this means one million pages. At one centimeter per 100 pages, this is 10,000 centimeters or 100 meters—a stack of paper whose height is about equal to the length of a football field.

Activity 8 All possible 2-letter words in DNA:

AA	AC	AG	AT
CA	CC	CG	CT
GA	GC	GG	GT
TA	TC	TG	TT
	=	**16**	

Activity 9: All possible 3-letter words in DNA:

AAA	AAC	AAG	AAT
ACA	ACC	ACG	ACT
AGA	AGC	AGG	AGT
ATA	ATC	ATG	ATT
CAA	CAC	CAG	CAT
CCA	CCC	CCG	CCT
CGA	CGC	CGG	CGT
CTA	CTC	CTG	CTT
GAA	GAC	GAG	GAT
GCA	GCC	GCG	GCT
GGA	GGC	GGG	GGT
GTA	GTC	GTG	GTT
TAA	TAC	TAG	TAT
TCA	TCC	TCG	TCT
TGA	TGC	TGG	TGT
TTA	TTC	TTG	TTT
		=	**64**

Activity 13 Possible DNA walk rules:

	step left for base	step right for base
1	A	C, G, T
2	C	G, T, A
3	G	A, C, T
4	T	A, C , G
5	A, C	G, T
6	C, G	T, A
7	A, T	C ,G

Plus 7 more that interchange last two columns.

Activity 4.4.18 Scaling exponents for profiles 4.9- 4.11.

Profile	Exponent
1	0
2	1
3	What did you find for your profile?

Activity 23 For a 25-letter alphabet we can find **33,554,431** different walk rules. If we include all walk rules, which lead to exact mirror images of the profiles, we even obtain twice as many walk rules. If we include punctuation symbols in our alphabet, find for a 32-letter alphabet **2,147,483,647** (4,294,967,294 respectively) walk rules.[a]

---

[a]The general rule is that for a n-letter alphabet where n is at least 2, you can find $2^{n-1} - 1 = (2 \cdot (2^{n-1} - 1))$ possible walk rules.

Activity 24 These are some of our ideas to define the activity:

- Define a walk rule. Examples:

  - Use the right half of the keyboard to step up, and the other half to step down.
    Or
  - Use the first half of the alphabet to step up, and the second half to step down.

  These are just two ideas, you know there are a lot more.

- Use different kinds of text. You can use texts which you have already analyzed with the programs **Zipf**.

- Construct the landscapes. Can you find differences in the structure of the landscapes for different texts?

- Calculate the scaling exponent for the texts.

- Write your results in a table.

- Try a different walk rule.

- Repeat the last three steps.

- Compare your results. Do you observe noticable changes after choosing a new walk rule?

Appendix B

Log-Log-paper

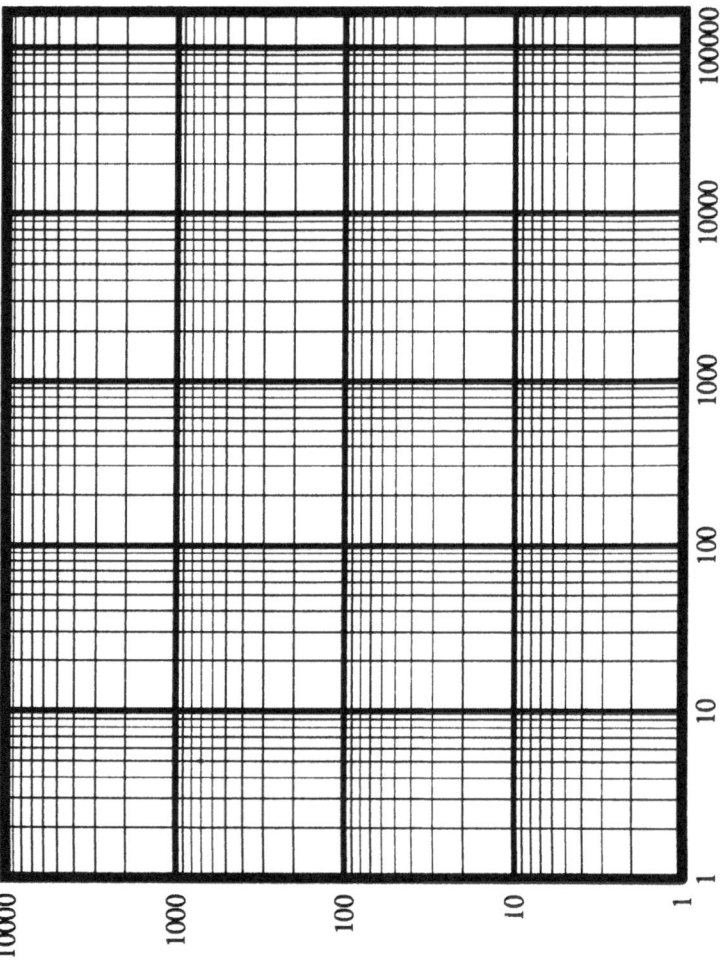

Figure B.1: Log-Log Paper. Photocopy this page and use copies.

Appendix C

This and That

C.1 Translating codons

Codons and the amino acids they code for.

Compare Box 5

First Position	Second Position				Third Position
	T	C	A	G	
T	Phenylalanine	Serine	Tyrosine	Cysteine	T
	Phenylalanine	Serine	Tyrosine	Cysteine	C
	Leucine	Serine	Stop	Stop	G
	Leucine	Serine	Stop	Tryptophan	G
C	Leucine	Proline	Histidine	Arginine	T
	Leucine	Proline	Histidine	Arginine	C
	Leucine	Proline	Glutamine	Arginine	A
	Leucine	Proline	Glutamine	Arginine	G
A	Isoleucine	Threonine	Asparagine	Serine	T
	Isoleucine	Threonine	Asparagine	Serine	C
	Isoleucine	Threonine	Lysine	Arginine	A
	Methionine	Threonine	Lysine	Arginine	G
G	Valine	Alanine	Aspartic acid	Glycine	T
	Valine	Alanine	Aspartic acid	Glycine	C
	Valine	Alanine	Glutamic acid	Glycine	A
	Valine	Alanine	Glutamic acid	Glycine	G

Table C.1: Stop means that these codons are stop codons. They are termination signal for the translation of the code. ATG is part of the inition signal, but unfortunately it codes also for the amino acide *Methionines*. Its code is ambiguous and does not help to identify coding regions. Do you see that the base at the third position does not play a role as important as the leading two bases because often you can change the base at the third position in a codon without changing the amino acid.

C.1.1 Translation of the poem Nell

Translation of **Nell**

Your purple rose in your brilliant sun,
Oh June, sparkles as if intoxicated,
Bend toward me, too, your golden cup:
My heart and your rose are alike.
Under the soft shelter of shady boughs
Sound a voluptuous sigh;
And turtle doves coo in the spreading wood,
Oh my heart, their amorous lament.
How sweet is your pearl in the flaming sky,
Star of the pensive night!
The singing sea, along the shore,
Will silence its everlasting murmur,
'Ere in my heart, dear love,
Oh Nell, your image will cease to bloom!

Contents

Chapter 5. Spin Glasses and
Neural Networks

Chapter 5

Spin Glasses and Neural Networks

5.1   Introduction

In previous manuals you have learned that order can grow out of disorder and randomness. You may have noticed a remarkable property of disordered systems—the larger number of elements that comprise a disordered system or that take part in disordered dynamics, the more dependable are laws that govern the "new order." The recognition of these laws is a triumph of modern science. In this workbook we apply these wonderful results to an even more awesome question: "How does the brain recognize patterns?" Strangely, the answer to this question begins with the tangle of molecules in glass.

QUESTION TO THINK ABOUT: Try to find a glassy marble, a glassy pearl, or any solid piece of everyday glass. Then take an ice cube and look carefully at the two objects. Can you say that they are similar, or do you find that they are made of quite different materials?

The Irish physicist John Tyndall (1820–1893) wrote that ice has *"the same relation to glass that orchestral harmony does to the cries of a market place; the ice is music, the glass is noise; the ice is order, the glass is confusion."* With the equipment available at the time, Tyndall could not verify how right he was about the molecular structure of ice and glass (see Figs. 1 and 2). Nowadays we know that, on the molecular scale, elements of ice are connected in an ordered network—a crystal lattice, whereas elements of a glass are connected in an apparently disordered manner. On the other hand, solid pieces of ice and glass may look quite similar; both materials can allow visible light to pass through. (We hope you came to the same conclusion when we asked you to think about it).

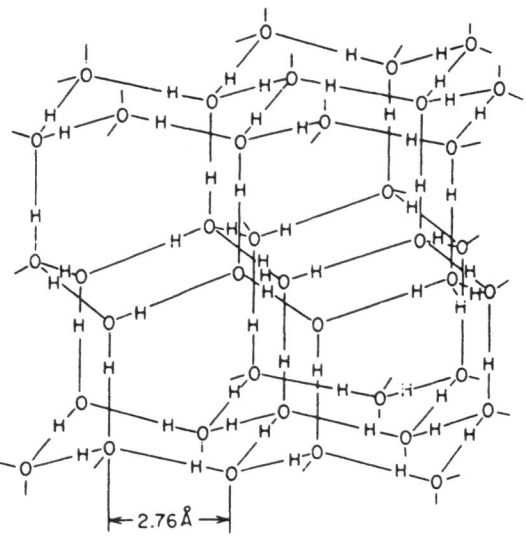

Figure 5.1: The structure of ordinary (H_2O) ice crystals [according to: S. N. Vino-gradov and R. H. Linell, *Hydrogen Bonding* (Van Nostrand, New York, 1971), p. 201].

Tyndall wrote complimentary words about ice in the last century. Today we can add that the disorder which occurs within the glassy state of matter has its own merits. One specific kind of glassy state that occurs in certain new magnetic materials named **spin glasses** appears to be particularly useful. We can think of the molecules in spin glasses as little bar magnets. In trying to describe how these little magnets line up with one another, workers in the field discovered that the way these systems change the the lining up of magnets is quite similar to one important aspect of the brain functioning, that is, to the so–called *associative memory activity*. But the story does not end at recording the similarity between the two apparently different structures (brain function and properties of disordered magnets). Knowing the dynamics of spin glasses one can make a program that allows the computer *to learn* a number of patterns (pictures of human faces and cars, for instance) and *to correctly classify* (recognize) a new pattern when it is loaded into its "brain." Once you understand this **pattern recognition** process, you can play with it and do further research on your own. To this end, we start with models of spin glasses and show later that the same models can mimic the associative memory activity of the brain.

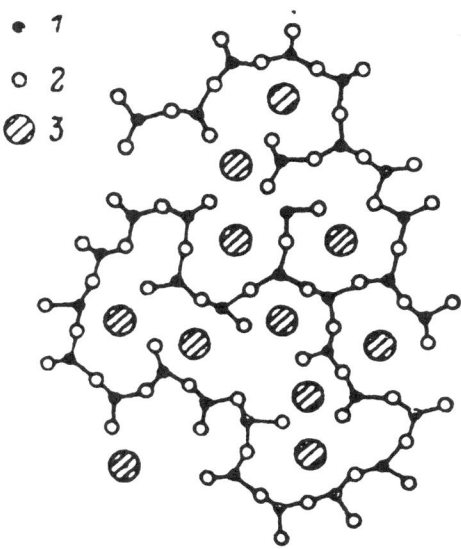

Figure 5.2: Schematic bonding of elements in layers of ordinary glass [according to: B. E. Warren, *J. Amer. Ceram. Soc.* **24**, 256 (1941)]. The small solid circles (•) represent the silicon (Si) atoms, the small open circles (o) represent oxygen (O) atoms, and finally the crosshatched big circles represent sodium (Na) atoms.

Start with ordinary magnetic materials that have been known for more than a thousand years, and ask why do magnets attract (or repel) one another, and why do they attract (never repel) objects made of iron?

5.2   Magnetic Order

We may say that magnets that we come across in everyday life do what they do because they possess a very fine order within themselves. In fact, they lose their magnetic properties when this order is destroyed, for instance by heating.

QUESTION TO THINK ABOUT: What are the things that are ordered within magnets?

We cannot see directly the fine order that defines a magnet, but various physical experiments provide evidence that a magnet can be visualized as a collection of a large number of tiny magnets whose centers are fixed so that they can flip around. Of course, the tiny magnets are actually atoms; instead of speaking about the magnets, we can speak about spins of the atoms, or simply about spins. Spins can be represented by little arrows (vectors); when the majority of of the little arrows (the spins) are oriented in the same direction, the result is a large magnet of the kind we see every day (see Fig. 3).

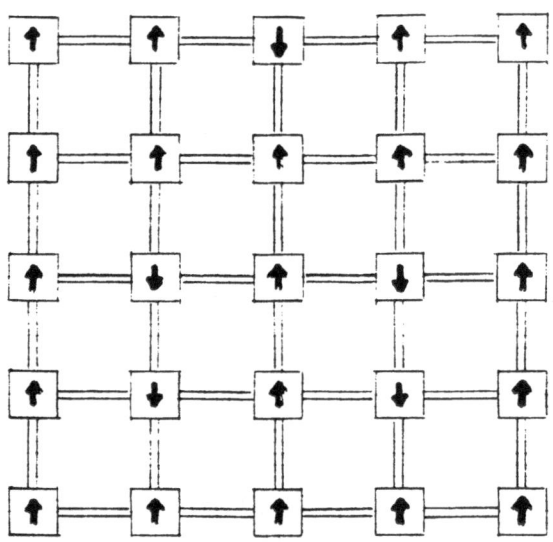

Figure 5.3: An ordered magnetic state on a square lattice (grid)—the majority of spins (bars with arrowheads) are upward oriented.

Can we be more precise in describing magnetic order? Can we make *a model* that relates magnetic order to the energy content?

Like toys, scientific *models* can be simple or complicated. Like toys, the best models are simple, durable, and unfailingly amusing. By these criteria, the best model of magnetic orderings is the **Ising model**, which we now describe.

5.3   Ising Model

In the Ising model, a spin at the lattice site i is assumed to have only two possible orientations—when it is upward oriented we say the that spin value is $S_i = +1$, whereas the value $S_i = -1$ means that the spin is oriented down. A pair of two spins, S_i and S_j, contributes an energy $-J_{ij}S_iS_j$ to the total energy E of the magnet. Here, J_{ij} is a constant whose value depends on the distance between the lattice sites i and j, and, of course, on the nature of magnetic material we look at.

A FEW MORE USEFUL WORDS: The constants J_{ij} are called **coupling constants**. A material is called **ferromagnetic** if all J_{ij} are positive, while a material with negative J_{ij} is called **antiferromagnetic**. In ferromagnetic materials, most of the spins are aligned n the same direction. In contrast, for antiferromagnets spins on adjacent sites tend to be oppositely-oriented. The standard magnets of everyday life are almost always ferromagnets, but never antiferromagnets. For instance, iron (Fe), nickel (Ni), and cobalt (Co) are ferromagnets, whereas chromium (Cr) and manganese (Mn) are antiferromagnets. Good everyday magnets (so–called permanent magnets) are made of iron alloys. The best–known are "alnico" alloys which, in addition to iron, contain <u>al</u>uminium, <u>ni</u>ckel, <u>co</u>balt, and even some copper.

5.3.1   Activity 1: Energy of Ferromagnets

Imagine a very simple magnet consisting of four spins that are arranged on corners of a square. The corners are enumerated 1, 2, 3, and 4, in the clockwise direction (see Figure 4). Besides, let us assume that all four bonds along the square sides are ferromagnetic with coupling constant equal to one, $J_{ij} = 1$, and that there are *no* spin interactions along the square diagonals (e.g., $J_{13} = 0$). In which of the following two cases does the magnet have a lower energy E: (a) when all four spins are upward oriented, (b) when three of them are upward oriented and one of them (the lower right one, for instance) is downward oriented? In answering this question, keep in mind that no strength parameter J_{ij} can change its value in going from case (a) to case (b), since the values of the parameters are fixed by the "nature" of the system. In contrast, when two spins are opposite, their interaction term $-J_{ij}S_iS_j$ takes on a positive value rather than a negative value. For the two cases we want to calculate the value of the expression:

$$E = (-J_{12}S_1S_2) + (-J_{23}S_2S_3) + (-J_{34}S_3S_4) + (-J_{41}S_4S_1)$$

For case (a), all spins aligned, we have the result:

$$(a)\quad E = -1 - 1 - 1 - 1 = \quad -4$$

whereas for case (b), spin 3 opposite to the others, we have the result:

$$(b)\quad E = -1 + 1 + 1 - 1 = \quad 0$$

Thus we see that energy in case (a) is negative, while in case (b) it is equal to zero. We can say that the simple Ising ferromagnet has lower energy when all its spins are upward oriented. However, you can imagine a third case (c) in which all spins are downward oriented and compare its energy with case (b). You can check that in the (c) case energy is again equal to −4. In general, the total energy E of an Ising magnet with all positive bonds ($J_{ij} > 0$), on a lattice consisting of a single or many squares, has the smallest value when all spins are oriented in the same direction.

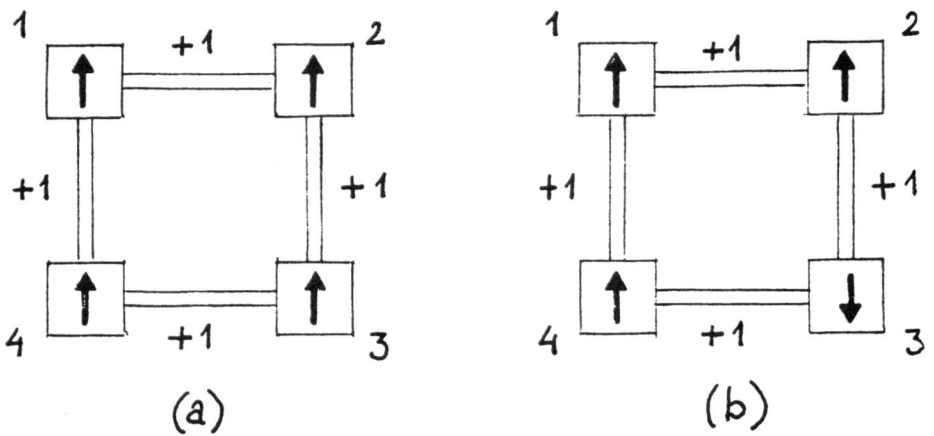

Figure 5.4: Two arrangements of spins of the Ising ferromagnet on a single square.

In the real world all systems tend to settle down into a state of the smallest (minimum) energy. In the case of the Ising model, the total energy consists of contributions of the type $-J_{ij}S_iS_j$ for each bond. If J_{ij} is positive and both spins are oriented in the same direction, then $-J_{ij}S_iS_j$ is a negative number and we say the i–j bond is **satisfied** (in the sense, that it brings the whole system closer to its lowest energy). In contrast, when J_{ij} is positive and the spins are oppositely oriented the energy contribution is a positive number, and we say that the bond i–j is **unsatisfied**, because this relative orientation makes the energy of the system higher. Finally, when there are many bonds in the Ising system, we use the shorthand notation

$$E = -\sum_{i,j} J_{ij}S_iS_j \; , \qquad (5.1)$$

to denote the total energy as the sum over all bonds in the system.

[For practice, write the previous equation for E in the form of the above shorthand notation. The only thing you have to do is to write two such expressions and indicate the range of values for the *index i* and the range of values for the *index j*, in the case of the ferromagnet depicted in Figure 4. Is it difficult?]

COMPUTER ACTIVITY 1: ENERGY LANDSCAPE — THE CASE OF A FERROMAGNET

[a] The first case we study is that of a ferromagnet, in which spins have the lowest interaction energy when they line up in the same direction. Move to the computer program called SPIN GLASS and choose the option "Bonds."

[b] You are asked to choose the size of the lattice (L), that is, the number of spins on the lattice ($L \times L$). By default $L = 5$ and there are 5×5 spins on the lattice. You can choose less or more, but for the time being we suggest you choose the default value and click on **Done**.

In the upper right part of the screen, the computer displays, among other things, a drawing of 25 spins depicted as small squares (compare with Figure 3), randomly colored yellow (indicating an up spin) or blue (down spin).

[c] Notice that there are no bonds between the "spins." To add these bonds, move to the **Model** menu and choose the command **FerroMagnet**. Then the system of 25 spins (small squares) acquire ferromagnetic bonds, colored blue to indicate that the interaction strength between any two adjacent spins i and j is positive, $J_{ij} > 0$. (In the next computer activity the antiferromagnetic bonds will be colored red.)

Your "playground" is one third of the screen, on the right, while your "score" is graphically recorded in the larger part of the screen at the left. The lower right corner of the screen is a kind of "template" which reminds you of he *unsatisfied* ($E = +1$) and *satisfied* ($E = -1$) bonds. Remember—the goal of the "game" is to arrange the spins on the 5×5 lattice so that as many bonds as possible are satisfied.

Start the game! You cannot change the color of the lattice bonds (since you have chosen the model to be a ferromagnet), but you can change the orientation of spins. Click on any of the yellow small squares and you change its color to blue, which means that you have changed its orientation from being $S_i = +1$ to $S_i = -1$. You can do the same with a blue square, changing the orientation form -1 (down) to $+1$ (up) orientation and its color to yellow. Moreover, you can do it with all 25 small squares (spins) as many times as you want. While you are flipping spins, you observe three changes—two below the main (top) lattice and a major change on the left part of the screen. First, just below the top lattice, you can see the energy (E) value that corresponds to the current arrangement (configuration) of spins and the minimum energy that the spin system has achieved during the time that you have been playing with it. Below these two lines, there is another lattice (grid) of the same size as the top one. "It is somehow incomplete!" you say? Yes, you are right. This lattice is comprised of only those bonds of the top lattice that are satisfied. Flip some spin on the top lattice and notice that the lower lattice either loses some bonds or gains some bonds, depending on whether the spin flipping made more satisfied bonds or it has destroyed satisfied bonds that already existed. So, we can rephrase the goal of the game—one should try to make the lower lattice (composed of the black bonds) as complete as possible. In the case of a ferromagnet, this lattice can be made 100% complete. To achieve this, you should work only on its missing bonds, flipping spins at either end of such bonds.

Finally, we come to the left part of the screen, where you can see the graphical representation of the "score" of your activity. On the vertical white line there is a scale of possible energies of the spin system, while on the horizontal white line no scale is visible. However, the horizontal line is in fact divided into a large number of equal little segments such that each segment corresponds to one flip of a spin, starting at the left of the graph. When the "game" starts, and you start to flip spins, the program SPIN GLASS starts to draw a curve in stepwise manner. The beginning of the curve corresponds to the energy of the initial configuration of the spin system, before you began flipping spins. As you proceed to flip spins, the curve goes up and down, indicating changes in the energy. What does it mean when the curve goes neither up nor down, but rather stays horizontal? This is not a hard question and we are sure that you can immediately provide the right answer. (The flat parts of the curve correspond to those flipping of spins that do not change the energy.)

When you look at the curve drawn in the left part of the screen, after a period of your activity, you may notice sharp peaks, valleys, passes, plateaus, valleys, gorges and many other similarities with a profile of a mountain landscape. That is the reason why we say that this curve depicts *the energy landscape* of our spin system. Accordingly, the goal of the "game" is to find (by flipping spins) the lowest possible point of the landscape. When you reach it, the lattice comprised of satisfied bonds will be completed and all spins will be oriented in the same direction— all small squares will be colored in the same way (either yellow or blue). Now, read the two lines below the top lattice. The value that you see is the lowest possible energy of the ferromagnet studied.

You may be tired, but we urge you to flip *all* spins in the opposite direction (change their color) to find the new value of E. Is it the same as in the previous case? How would you express your finding?

[d] If you are eager to continue the computer activity, and you want to confirm your findings, move to the menu **File** and choose the command **Change size**. Then you can enlarge (or even decrease) the size of the lattice you have studied so far, and repeat the most interesting parts of the foregoing activity, working now on the new lattice. Or choose **Quit** from the **File** menu and continue your study later. We believe you will like what comes next.

5.3.2 Activity 2: Energy of Antiferromagnets

In this activity, we start with a single square and the Ising spins coupled by antiferromagnetic bonds of unit magnitude; that is, we assume $J_{ij} = -1$ for all bonds (see Figure 5). Next, we compare two of many possible spin arrangements: (a) all spins are oriented in the same direction, and (b) no two nearest neighboring spins are oriented in the same direction. We pose the same question as in the previous Activity: Which of the two spin arrangements have smaller energy? It is wise to make the simple calculation:

$$E = (-J_{12}S_1S_2) + (-J_{23}S_2S_3) + (-J_{34}S_3S_4) + (-J_{41}S_4S_1)$$

For case (a), all spins oriented in the same direction

$$E = +1 + 1 + 1 + 1 = +4,$$

whereas for case (b), neighboring spins oriented in opposite directions

$$E = -1 - 1 - 1 - 1 = -4.$$

Here we have assumed that no interaction parameter J_{ij} can change its value in going from the case (a) to the case (b), since the values of the parameters are fixed by the "nature" of the system. Hence, the last two lines of the calculation tell us that the arrangement with *antiparallel* neighboring spins has smaller energy. The following computer activity should lead you to the general conclusion that, when all bonds on a square lattice are antiferromagnetic ($J_{ij} < 0$), the total energy E of the Ising system of spins is the smallest for the arrangement in which no two spins connected by a single lattice bond are parallel. What does this arrangement look like?

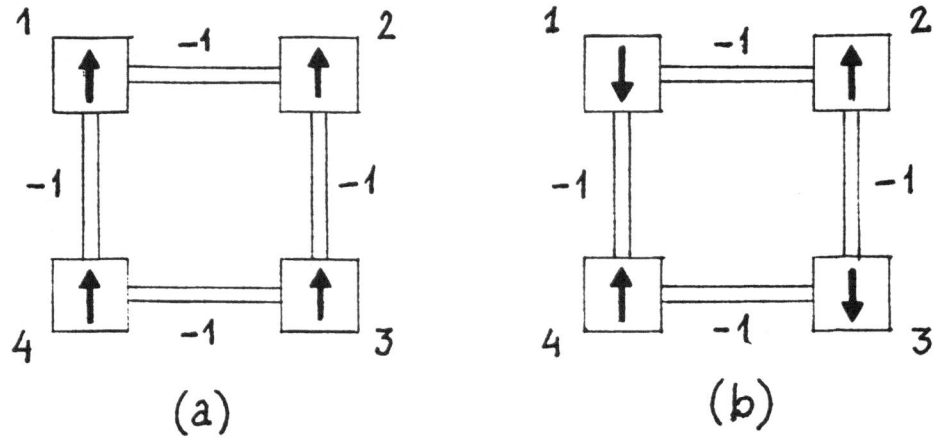

Figure 5.5: Two arrangements of spins of the Ising antiferromagnet on a single square.

In order to visualize a larger antiferromagnetic lattice with the smallest energy (before starting the computer activity), depict the state on a piece of paper, using a square lattice of the size presented in Figure 3, that is, on the lattice with 25 sites. Start with the central site, drawing there an upward-oriented arrow to represent spin and four downward-oriented arrows to represent spins on its four nearest neighbors. Then, in the next step, surround each of the latter four spins with three upward oriented nearest neighbors, and so on up to the last 25th spin. Eventually, you should have the entire lattice covered with the expected arrangement of spins. Now, you can color in yellow (or in black) each small square that contains upward oriented spin, and in blue (or in white) those small squares that contain spins oriented down. Compare the antiferromagnetic state of smallest energy to a checkerboard.

COMPUTER ACTIVITY 2: ENERGY LANDSCAPE — THE CASE OF A ANTIFERROMAGNET

[a] Now redo the 5×5 antiferromagnetic lattice with computer. Move to the program SPIN GLASS and choose $L = 5$ from the **Bonds** menu, to play again with 25 spins.

[b] Now choose **AntiFerromagnet** from the **Model** menu. From the previous computer activity, you remember that your "playground" is the top lattice in the right part of the screen. You notice that the bonds of the lattice are now colored red to indicate that the interaction

strengths between any two adjacent spins i and j is negative, $J_{ij} < 0$. Now start to flip spins trying to achieve the goal, that is, to make all bonds *satisfied*. Is it possible? Your guide should be the lower lattice— it should be made as complete as possible. Without a missing bond? Yes, without a missing bond on the lower lattice! You can follow your score in the two lines between the upper and lower lattices and on the graph in the left part of the screen.

As you flip the spins, follow the way the *energy landscape curve* develops at the left of the screen, and notice that its peaks and valleys now correspond to different arrangements (configurations) of spins than in the case of a ferromagnet. But the goal of the game is the same— by flipping spins you should find the lowest "valley" of the landscape. When you reach it, the lower lattice (composed of the satisfied bonds) should be complete, while the top lattice should look like a checker- board, comprising alternating yellow and blue small squares (up and down spins). The two lines between the two lattices tell you the value of the corresponding energy minimum. When you learn this value, make your discovery complete by turning all spins into the opposite directions, which is equivalent to changing the color of the small yellow squares into blue and *vice versa*. While you are doing this, the energy landscape curve rises and falls, but eventually it reaches the same lowest "valley" it had before. Are the energy values recorded between the two lattice as low as they have been before? Compare your finding for the antiferromagnet with the similar finding in the case of the ferromagnet.

A two-word phrase that describes your findings: The state of a phys- ical system in which it has the smallest energy is called the **ground state**. Hence, we can say that an Ising ferromagnet is in its ground state when all its spins are parallel, whereas an Ising antiferromagnet is in its ground state when it has no parallel spins on nearest neighboring lattice sites (in other words, when each spin is surrounded by differently oriented spins).

5.4 Magnetic Disorder

A ferromagnet that is properly described by the Ising model can be found in its ground state only at the absolute zero temperature (that is, at -273°C). Then its spins display maximal order (all are oriented in the same direction). When temperature increases, spins jiggle around and start to flip randomly; the order decreases. Order vanishes completely at a certain critical temperature (the so-called *Curie point*). At and above this critical temperature one half of the spins, on average, are oriented in one direction and the other half are oriented in the opposite direction. It is important to note that any particular spin, at a given lattice site, can be at one moment upward oriented, and at another moment it can be downward oriented as spins participate in the thermal motion.

SOME DATA: The critical temperatures (the Curie points) T_c of iron (Fe) and cobalt (Co) are very high. Thus, $T_c = 770^0$C for iron, and $T_c = 1115^0$C for cobalt. On the other hand, one of the major ores of iron—the black mineral **magnetite**, whose chemical name is ferrosferric oxide (Fe_3O_4)—has a bit lower critical temperature, $T_c = 585^0$C. Certain varieties of magnetite, known as **lodestone**, occur as natural magnets and were used as compasses in the ancient world.

At temperatures between absolute zero and the Curie point, the spins of a ferromagnet are not all parallel. How many of them are parallel? It depends how far the temperature is from absolute zero and how close it is to the Curie point. The percentage of spins that are, on average, antiparallel to the original preferred direction, increases gradually as temperature increases. Thus, we can say that the higher the temperature, the larger the disorder it induces.

QUESTION TO THINK ABOUT: Is there something else, besides the temperature, that can induce magnetic disorder?

This is not an easy question. Here is the answer: If we can somehow eliminate spins at certain lattice sites we can diminish magnetic order from the start (that is, even at the absolute zero of temperature). How can we eliminate spins from some lattice sites? In nature, and in laboratories, this happens by substituting nonmagnetic atoms for magnetic atoms. The substitution can occur in various ways and with different ratios of magnetic to nonmagnetic atoms. It can be accomplished under various experimental conditions, producing materials with different structural and magnetic properties. For instance, in certain materials so few magnetic atoms are present that no bulk magnetic order can exist even at absolute zero temperature.

> AN ANALOGY: In painting, you can mix primary colors (red, yellow, and blue) to produce either secondary or intermediary colors, and by adding white and black color you can produce various tints and tones. Similarly, mixing of magnetic atoms with one or two kinds of nonmagnetic atoms is an art that belongs to **materials science**. Products of this art can be very appealing.

5.5   Spin Glasses

Spin glasses are modern products of materials science. A spin glass is typically a dilute (substitutional) alloy of a small percentage (between 0.05 and 15%) of a magnetic metal, such as iron or manganese, in a host noble-metal, such as copper (Cu) or gold (Au). In this case, magnetic atoms are distributed randomly in the host crystal lattice (see Figure 6). If we try to visualize a collection of lines that connect these atoms we get a network that is typical for a glassy state. More importantly, magnetic interactions along the bonds between magnetic atoms can be either of the ferromagnetic or antiferromagnetic type. The net effect of such couplings between magnetic atoms is that certain thermal and dynamical properties of spin glasses are quite similar to some properties of ordinary glasses, and this was the reason for coining the phrase *spin glass*.

[NOTE: There are other types of spin glasses which contain higher concentrations of magnetic atoms, but both magnetic and nonmagnetic atoms are *randomly* distributed—neither of them constitute a crystal lattice. Hence, we say that the corresponding materials are **structurally amorphous** and in this way they are quite similar to ordinary glass. Examples of this kind of spin glasse are $GdAl_2$ and $(Fe_xMn_{(1-x)})_{0.75}(P_{16}B_6Al_3)_{0.25}$, where x is the relative concentration. In the first example there is one ferromagnetic atom of gadolinium (Gd) for every two nonmagnetic atoms of aluminium (Al). On the other hand, in the second example ferromagnetic iron (Fe) and antiferromagnetic manganese (Mn) comprise 75% of the material, and the remaining 25% constitute nonmagnetic phosphorus (P), boron (B), and aluminium atoms.]

The collection of spins in a spin glass is not capable of sustaining a magnetic order of the type we have spoken about earlier in this chapter.

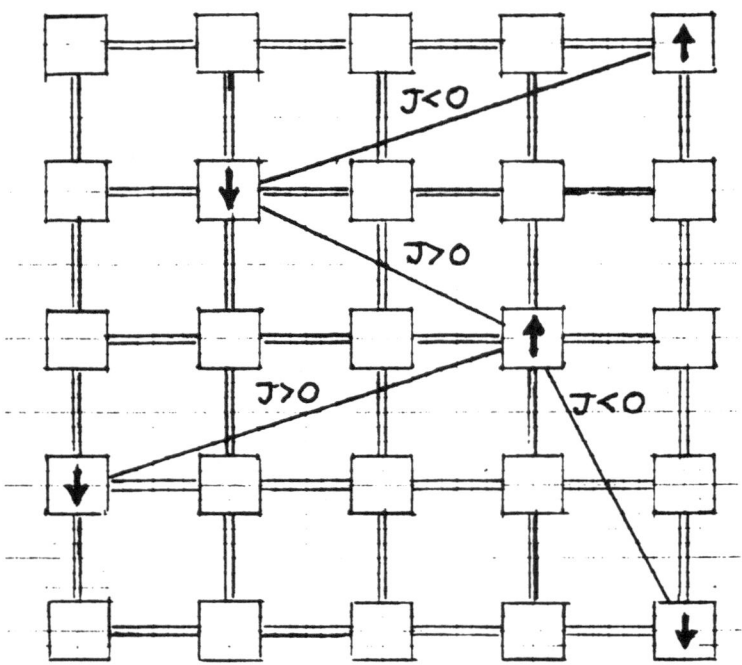

Figure 5.6: A random distribution of magnetic atoms in a spin glass. Because of the presence of intervening nonmagnetic atoms, the couplings between spins can be ferromagnetic ($J > 0$) as well as antiferromagnetic ($J < 0$). The small empty squares represent sites of nonmagnetic atoms.

For instance, you cannot use a spin glass to collect spilled nails and pins. However, spin glasses have other useful properties in the realm of materials science. Besides, couplings of spins in a spin glass exhibit some new interesting features. But, before saying anything more about the spin glasses, we ask two simple questions.

1. Have you ever been frustrated?

2. Can you describe what it feels like to be frustrated?

If your answer to the first question is "never," you are a lucky person. But more likely you have forgotten the situation when you were young and your mother told you "You can go outside!" whereas your father told you "You can*not* go outside!" Wasn't this frustrating? Or perhaps you have heard some of your friends say that they are frustrated in math classes. This can happen because math is full of rules and, if one does not practice enough, one cannot decide which rules to apply in a given situation. More generally, in a state of frustration we have an an impulse to do something that we are prevented from doing.

5.5.1 Activity 3: Frustrated Magnetic Bonds

Presence of both ferromagnetic and antiferromagnetic couplings in a system of spins can force some bonds to be frustrated. What does this mean? To see it, let us calculate total energy of four spins situated on the corners of a square, and let us assume that there are three antiferromagnetic and one ferromagnetic coupling, all of unit strength (see Figure 7). Furthermore, let us compare two configurations (arrangements) of spins that differ only in the orientation of the third spin (the one at the lower right corner). Following the rules we applied in our

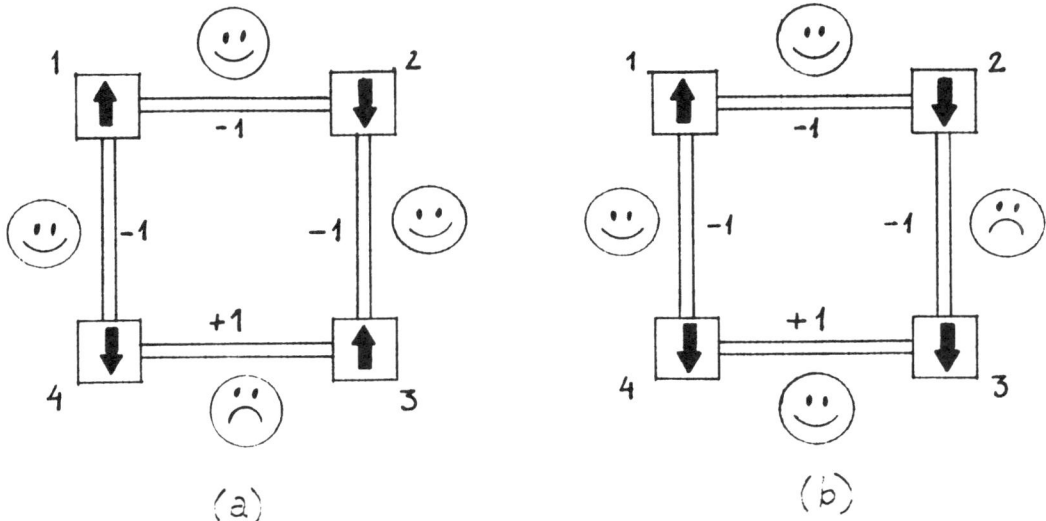

Figure 5.7: Two different arrangements of spins that have the same minimal energy $E = -2$ for the given set of spin couplings.

Activities 1 and 2, we find in case (a):

$$
\begin{aligned}
E &= (-J_{12}S_1S_2) + (-J_{23}S_2S_3) + (-J_{34}S_3S_4) + (-J_{41}S_4S_1) \\
&= -1 - 1 + 1 - 1 = -2,
\end{aligned}
$$

and, interestingly enough, in case (b) we get the same result:

$$
\begin{aligned}
E &= (-J_{12}S_1S_2) + (-J_{23}S_2S_3) + (-J_{34}S_3S_4) + (-J_{41}S_4S_1) \\
&= -1 + 1 - 1 - 1 = -2.
\end{aligned}
$$

Hence we see that both configurations of spins have the same energy $E = -2$. Let us specify their similarities and differences. It is not a difficult task! First, we notice that in case (a) all three antiferromagnetic bonds have differently oriented spins on their ends, exactly as should be expected in a ground state configuration. However, the ferromagnetic bond (the base of the square) also has differently oriented spins on its ends, which is not typical for a ferromagnetic coupling. For this reason, we say that the three antiferromagnetic bonds are *satisfied*, while the ferromagnetic bond is *frustrated*. On the other hand, in case (b) two antiferromagnetic bonds and one ferromagnetic are satisfied, whereas the third antiferromagnetic (the right vertical side) is *frustrated* since spins on its ends are pointed in the same direction.

By exploring this set of couplings you can discover, in addition to the cases (a) and (b) depicted in Figure 7, six more configurations of spins that have $E = -2$. For instance, start with the (a) case and turn up the fourth spin, and then calculate the resulting energy.

Now, a new question: For the same set of couplings, are there spin configurations that have energy larger than -2? There are eight such configurations, and all of them have $E = +2$. You can find them yourself! We are sure you can do it. (To begin with, take the (b) case and turn up the fourth spin.)

Therefore, we may draw the conclusion: *Systems with frustrated bonds have many (more than two) states with minimal energy.* Indeed, in the case of spin couplings depicted in Fig. 7 there are eight states with $E = -2$. However, in science, it is extremely dangerous to take for granted a conclusion drawn from one particular example. For this reason, we suggest that you use the following activity with the computer program SPIN GLASS and convince yourself that this conclusion is correct for many more cases.

COMPUTER ACTIVITY 3: ENERGY LANDSCAPE — THE CASE OF RANDOM MAGNETIC BONDS

[a] Start this activity with a very small system. Start the program SPIN GLASS, click on "Bonds," and choose $L = 3$. Now, with the new image on the screen, move to the menu item **Model** and select the command **Random**. On your "playground' (the upper part on the right third of the screen), you see a lattice with 3×3 spins and with randomly assigned ferromagnetic (blue) and antiferromagnetic (red) bonds between pairs of adjacent spins. This distribution of bonds (blue and red) should be regarded as fixed by the 'nature" of the spin system under study. In other words, the spins can be flipped, but the bonds, for the given lattice, cannot change colors.

[b] As in the previous two computer activities, the goal of the game is to find the configuration (arrangement) of spins, on the given lattice,

that has the lowest energy. The game is played in the same way as before—by flipping the spins on the upper lattice. However, to achieve the goal is now a bit harder. Nevertheless, we are sure that you are not going to give up. Moreover, you will be able to develop guidelines for finding the spin configuration with the lowest energy. By the way, if you run out of space in the left part of the screen, select the command **WrapPlot** from the **Options** menu. This tells the computer to start drawing the continuation of the energy landscape curve above the previous one.

In the case of the previous two activities the guidelines turned out to be simple: for a ferromagnet, all spins should be oriented in the same direction, and, for an antiferromagnet, alternate spins should be oriented up and down. In the case of random bonds, the guidelines are slightly more complicated: if we focus our attention on the 4 squares made up of 4 spins and 4 bonds (all being elements of the upper lattice), we shall say that *a square is frustrated* if any of its four bonds is frustrated. Then, after a bit of inspection, we see that squares with an odd number of antiferromagnetic bonds are frustrated, while those with an even number of antiferromagnetic bonds are satisfied. Finally, to achieve the goal of the game, we learn that the best one can do for a frustrated square is to make satisfied 3 out of 4 bonds, whereas, for a square that is not frustrated, one can make all bonds satisfied.

We hope that you have discovered the same guidelines described above. Let us confirm them—move to the menu item **File** and select the command **Change size**. Choose a larger lattice, say $L = 6$. Again, flip spins to find the lowest energy configuration. Follow the guidelines—take each square and see if it is frustrated or not, then try to satisfy as many of its four bonds as possible. Do this for all squares that constitute the upper lattice. In doing so you may come across two adjacent frustrated squares. Which bonds are you going to make satisfied in such a case so as to achieve the lowest energy?

[c] As regards the graphical representation of your score (on the left part of the screen), you may observe that the "energy landscape" contains more varieties than in the two previous activities. As you flip spins and the energy landscape curve develops, there appear many hills and valleys. Sometimes you can find yourself at the bottom of a valley that does not correspond to the lowest energy state, and then you have to climb out of this valley (that is, to increase the energy of the system, by flipping some spins) in order to get into a deeper valley corresponding to an even lower energy, and so on. In other words, there are valleys of various depths separated by hills or mountains, indicating

that there are many states with local energy minima. *This property is typical for disordered and frustrated systems and leads to their glassy behavior.*

[d] As mentioned earlier, we know, from everyday experience, that systems in nature tend to change so as to achieve a state of minimal energy. This is true for a mechanical as well as for other kinds of system, including spin glasses. Once a system is "let alone," with no external influences, it evolves to a state with minimal energy. The way it evolves to such a state depends on the nature of the system. Similarly, the time needed for the system to evolve to a state of minimal energy depends on the *dynamical rules* that are set (by the laws of nature) within the system. Finally, in the case of many states with minimal energy, the final state in which the system settles depends on the state it starts from. We can say that each state with minimal energy has its **domain of attraction**, which is in fact a set of all other states from which the system evolves to the one in question. In terms of the energy landscape and using a mechanical analogy, a domain of attraction comprises all local hills and mountains that surround a valley and from which a ball rolls down to the bottom of the valley. Like a portion of a golf course, the domain of attraction can be compared to all those points from which a golf ball will naturally roll toward the given hole.

A thought experiment: Next time you have difficulty remembering a name—for instance the name of a friend you see infrequently—try to remember a specific situation in which you heard or used the name, such as an experience you shared with your friend. In other words, think of as many relevant details surrounding the actual use of the name as you can. During this process, you may suddenly recognize some detail that helps you recall the friend's name. Thus, we can say that state of your mind has been in the *domain of attraction* of the name and eventually you got it. This act of recalling is called **association**, and this kind of memory is called **associative memory**. Later in this project we will use a computer program that mimics associative memory.

COMPUTER ACTIVITY 4: ENERGY LANDSCAPE — COMPARISON OF THE THREE MODELS

[a] In this activity we do not flip spins manually as we have been doing so far. Let us fix the system size at a larger value, say $L = 11$, and let the computer do the flipping and looking for the lowest energy states. How can it do this? There is an option built in the program SPIN GLASS that you may have already discovered. First, move to the menu

bar and select **FerroMagnet** from the **Model** menu. Now, go to the **Options** menu and select **Evolve**. The computer starts to flip spins on the lattice so that the energy of the spin system (model) decreases continuously. You can follow this process by observing the energy landscape curve on the left. The program has been designed to change the arrangement of spins according to certain *rules*. These rules are chosen so as to mimic the tendency of real systems to evolve towards an arrangement (configuration) of minimum energy. In other words, the computer flips a spin only if that results in decreasing the overall energy of the system. Therefore, select the option **Evolve** several time, until the spin configuration does not change any more. At this point, the spin system has attained the lowest possible energy.

How can we verify the preceding statement? Simply select **FlipRandomly** from the **Options** menu. The computer select spins randomly and flips them. So what? If the system was at a true minimum, the energy must increase by random flipping You may note the increase in energy as given in the line below the upper lattice, and also by following the energy landscape curve. To make it more convincing, choose **FlipRandomly** a couple of times. Thus you take the system out of its minimum energy state. Now, try the option **Evolve** again. Is the new energy minimum configuration the same as the one before (i.e., before you used the option **FlipRandomly**)?

[b] Go back to the menu bar and choose **Random** from the **Model** menu. Now repeat the procedure performed in the previous section of this computer activity. That is, choose **Evolve** several times to find a minimum energy, then disturb arrangement this by selecting **Flip Randomly** a few times, finally, choosing **Evolve** a few more times to find the energy minimum again. Do you get the same spin configuration for the energy minimum the second time as the first? Probably not. Can you explain why?

In the case of random or frustrated systems, there are many configurations with the same minimum energy. Thus, which configuration with minimum energy the system evolves to depends crucially on the configuration from which it starts to evolve. Once it reaches a state with minimum energy and the **FlipRandomly** command has been used, the system is unlikely to return to the same initial higher-energy state. Consequently, it is also unlikely that the new **Evolve** action will bring it back to the same minimum energy state as before.

[c] Finally, let us compare the energy landscape curves for the three different model systems—for a ferromagnet, an antiferromagnet, and a

spin system with random (mixed ferromagnetic and antiferromagnetic) bonds. Let us keep the system size fixed at $L = 11$. First, look at a typical ferromagnetic landscape. Choose **FerroMagnet** from the **Model** menu, then choose **Sweep** from the **Options** menu. This choice makes the computer flip spins of the system in such a way that they go from all up (yellow) to all down (blue). Keep selecting **Sweep** until the system has gone through the state with all (11×11) spins up (yellow) twice and through the state with all (11×11) spins down (blue) also twice. In doing this, you should observe four energy minima, that is, four deep valleys in the energy landscape curve. The bottoms (the four identical lowest "altitudes') of these deep valleys correspond to the *ferromagnetic ground states*, which can be either with all spins pointed up (yellow) or with all spins pointed down (blue).

To make comparison with the antiferromagnetic case, select **Anti-Ferromagnet** from the **Model** menu. Again select **Sweep** from the **Options** menu, and repeat the whole procedure done in the preceding case. When you look at the new energy landscape curve and at the upper right lattice, you see that the spin configurations that corresponded to energy minima for the ferromagnet (locations of the lowest "ferromagnetic" altitudes) are the spin states that correspond to top of the hills (the highest "altitude" points) for the antiferromagnet. Isn't this a sufficient reason to call the latter system *anti*ferromagnet?

As the last and most important part of the current computer activity, select **Random** from the **Model** menu, and repeat the **Sweep** procedure done in the ferromagnetic and antiferromagnetic cases. You see quite a different landscape curve, with many valleys that are not very deep. Can you explain this new feature? How is the multitude of valleys related to the fact that the random-bond magnet has many more energy minimum states than the ferromagnet, or the antiferromagnet? The deepness of the valleys is related to the fact that in the case of the random magnet all bonds cannot be satisfied (in contrast to the ferromagnetic and antiferromagnetic cases), and therefore the values of the corresponding energy minima result from summing some negative contributions along with a number of positive contributions that spring from the frustrated bonds.

5.6 Hopfield Model

In 1982 scientist John J. Hopfield introduced a mathematical model of associative memory which turned out to be the same model proposed a few years earlier by physicists David Sherrington and Scott Kirkpartick to describe states of spin glasses. The Hopfield model has become a major stepping–stone in scientific understanding of the way our brain works.

Is such a model possible? Is it complicated? Yes, it can be complicated, but by now you should be able to appreciate the basic concepts and enjoy their applications. Indeed, the mathematical form of the Hopfield model

$$E = -\sum_{i,j} J_{ij} S_i S_j \ , \tag{5.2}$$

is similar to the form we used earlier in this chapter (see, for instance, Activity 1). However, there is a new feature. Now it is assumed that there is a bond between *every* two spins (not only between adjacent spins), that is, i and j can be any two sites of the underlying lattice, no matter how far apart they are from one another. Besides, the spin interactions J_{ij} are *randomly* distributed on the lattice, so some couplings can be ferromagnetic ($J_{ij} > 0$), while others are antiferromagnetic ($J_{ij} < 0$). Finally, whatever the coupling is between two spins we assume it is symmetric: $J_{ij} = J_{ji}$.

The most interesting property of the Hopfield model is that it may have many spin configurations with local minimal energy (with many frustrated bonds). Each of these configurations has its own *domain of attraction*. That is, for each configuration there are similar states of the spin system that have slightly larger energies. Thus, if the energy minima correspond to some interesting patterns, and if there are simple dynamical rules of retrieving such states, the Hopfield model may serve as a good model of **pattern recognition**.

5.6.1 Activity 4: Ising Patterns

We start this activity with the simple question: How can a state of the Ising spin system represent a pattern? To answer this question, take the example given in Figure 3 and enlarge all boxes that contain spins until the boxes touch each other. Then, color black those boxes that contain spins oriented downward, and leave empty (or white) those boxes that contain upward oriented spins. The result should be the pattern given in Figure 8.

Of course, Figure 3 depicts one of many possible spin configurations. If we assume that it represents the Ising model with *all* couplings being ferromagnetic, then the particular configuration does not correspond to

Figure 5.8: The pattern that has one–to–one correspondence with the spin configuration given in Figure 3.

a state of minimal energy. However, it is not hard to imagine that there
is an Ising model with a *mixture* of ferromagnetic and antiferromagnetic
bonds distributed on the square lattice in such a way that the configuration in Figure 3, and accordingly the pattern of Figure 8, corresponds
to one of its states with minimal energy. In general, this can be achieved
by reducing the number of frustrated bonds. In the particular case of
Figure 3, this requires introducing some antiferromagnetic instead of
the assumed ferromagnetic bonds. Finding such a mixture of bonds we
call the **learning mode** of the Hopfield model. It is hard to do it "by
hand," so let the computer do the work for you.

COMPUTER ACTIVITY 5: LEARNING MODE — FINDING PROPER BONDS FOR A GIVEN PATTERN

We have learned that random systems have many possible states with
energy minima, and the state to which the system evolves depends
on the initial spin configuration. In general, the system evolves to
the closest minimal energy state. (In an analogy with the mountain
landscape, a stone or snowball rolls down into the nearest valley).

However, in the previous activities, all the minimal energy states were predetermined (fixed) by the given type of bonds, which remained unchanged for each experiment. More precisely, a bond configuration was chosen and then the corresponding spin configuration found that made the energy of the system minimal. The *learning mode* of the Hopfield model proceeds in an opposite sense—instead of choosing the bond configuration, we first choose a spin configuration, or *pattern*, and then invoke the computer program to deduce the corresponding bonds between spins such that the pattern chosen becomes an energy minimum of the spin system. This means that later on, given an arbitrary initial state, the system may evolve towards the state we have chosen at the beginning. In other words, the system has *learned* the chosen spin state, and, we may say that the computer "remembers" the corresponding pattern. In short, *the process of learning consists in assigning magnetic bonds in such a way that the chosen pattern becomes a state of minimal energy.*

The method described allows the computer that runs the SPIN GLASS program to learn more than one pattern, each new pattern corresponding to a spin state with a local energy minimum. In doing so, the computer program modifies each magnetic bond to have a numerical value determined by all the patterns memorized thus far. For example, if in the first pattern the spins S_1 and S_2 have the same orientation, then it is favorable to choose the ferromagnetic coupling between them, that is, $J_{12} = +1$. However, it may happen that in the next (the second) pattern, the spins S_1 and S_2 have opposite orientation, and then it is favorable to choose the antiferromagnetic coupling $J_{12} = -1$. Consequently, after the two patterns are learned, the program makes J_{12} equal to zero, and no bond is assigned between spins S_1 and S_2 (if the system does not learn any more pattern besides the first two). In contrast, two consecutive ferromagnetic bonds produce a ferromagnetic bond $J = +1$, and two consecutive antiferromagnetic bonds produce an antiferromagnetic bond $J = -1$.

Let's see what this looks like on the screen.

[a] Open the program SPIN GLASS and choose **HopfieldModel** under the **Models** menu. Within the new screen image, click on the option "Link." and accept the default value $L = 3$ for the lattice size. Your playground is again the right third of the screen.

[b] Flip the spins on the lattice in the top right corner so as to select a pattern. You may start with the letter "T" (by clicking on the spins in the first top row and on the two middle spins in the next two rows below).

[c] Then choose **Learn** from the **Options** menu. Pay attention to the result of the computer calculation—every spin of the lattice is connected to every other (not only to nearest neighbors as in previous computer activities). The new network of bonds represents the magnetic system whose energy minimal state is in fact the chosen pattern "T." To check this, flip a few spins and then select **Evolve** from the **Options** menu. You see that the energy landscape curve evolves to the previous low energy value of the pattern you asked the machine to memorize.

In the next action, you may choose a new pattern (say, the letter "C"), make the system "learn" it, and observe the new bond configuration. In this way you can create a spin system that "remembers" up to four patterns.

It is very important to observe the final configuration of bonds. Is this a complex network? Of course, it is not terribly complex, but you should imagine how it may look in the case of a lattice with a larger number of spins (for instance, in the case of the lattice from Figure 3 with 6×6 spins). In such cases individual bonds would hardly be discernible. For this reason, in the next two activities the network of bonds is not be shown on the screen.

COMPUTER ACTIVITY 6: PATTERN RECOGNITION — SIMPLE PATTERNS

[a] In the previous activity we watched the spin system 'learn" certain patterns by acquiring magnetic bonds (values of the interactions J_{ij}) such that the prescribed patterns are the energy minima of the system. In the present activity, the same concept is developed further, aiming towards the pattern recognition of real pictures.

[b] Open the SPIN GLASS program and choose the "Hopfield Model". Within the new screen image, click on the option "Hopfield Model," and then select a lattice size, say $L = 9$. Your playground is again the top part of the right third of the screen, where you see a lattice of 9×9 blocks (spins). Below the lattice there are four patterns that the system has learned beforehand. These patterns have been learned in the same way as was demonstrated to you in the previous Computer Activity, that is, by adjusting the values of the bonds between the spins. However, since the spin system is now larger, it gets too messy to show the corresponding bonds on the screen. So you can see only the patterns—in the upper row there is an "×" and a "+", while in

the lower row the two patterns are random. You may notice that the "playground" lattice on the top looks like a scrambled, or disfigured, version of the first given pattern.

Now, the game starts! Choose **Retrieve** from the **Options** menu. Keep selecting **Retrieve** until the image of the top 'playground" lattice does not change any more, which should mean that the corresponding spin system has reached a minimum energy state. Indeed, the option **Retrieve** basically performs the same function as the option **Evolve** in the Activity 4, that is it finds the closest state with minimal energy. In this case, it is probably the first memorized pattern, which allows us to say that the system "retrieved" the desired pattern. Why do we say so? Simply, because we started with the system being in a state that is close (but not identical) to pattern #1. In this way our game mimics the actual retrieval process of the brain: When we try to recall something, we start with some partial and sketchy information on the subject, say a face, and a successful retrieval process consists of an efficient bringing back to mind the entire picture.

If you like, you can start the retrieval process from a different initial state. To do this, click on the spins (blocks) of the top "playground" lattice to obtain an initial state of your choice, or use the command **FlipRandomly** to let the computer scramble the picture for you. You can keep going back to the option **FlipRandomly** to scramble (distort) the image by any desired amount. In doing this, you notice that the energy landscape curve starts from the bottom of a valley (that corresponds to the previously retrieved pattern) and develops further by acquiring a new hill, or a mountain.

Now, go back to **Retrieve** and see whether the system still retrieves the same pattern or a different one. This gives you an idea of how efficient the system is, that is, how far from a learned pattern you can start and still retrieve the original pattern. In other words, you see how large is the *domain of attraction* of the given pattern. Thus you can note that the system might sometimes also retrieve an image that is none of the four prescribed patterns. It is said that such an image corresponds to a "spurious state," which is usually a mixture of several patterns. Isn't this feature similar to something you might have experienced yourself? While trying to recall an incident or a face, don't you sometimes find that you have confused or combined two or more distinct memories?

[c] You can also draw your own patterns and let the system learn them. To this end, select the common **ClearScreen** from the **Options**

menu to clear a blank space within the "playground." Then click on squares until you get your own pattern composed of the yellow and blue blocks. Once you have finished your pattern, choose the command **Learn** and the system *learns* it. Every time the system learns a pattern, you notice that it removes one of the random patterns and places yours instead, whereas "×" and a "+" patterns always stay in the memory of this artificial "brain." Due to a lack of additional space on the screen, the number of patterns remembered by the system is kept constant at 4. Also, once a new pattern is learned, the computer automatically starts a new plot of the energy landscape curve. This is because (as we found in the previous activities) each new pattern changes the bonds of the system and thus the whole energy landscape must change. Finally, when your pattern is memorized, you can again scramble it using the **FlipRandomly** command, and then choose **Retrieve** to see if the system can "recall" it.

In playing with the Hopfield model on the computer, you will observe that frequently when you start from some random pattern similar to, but not exactly the same as, the learned random (or regular) pattern, the pattern that is finally retrieved may have certain "defects." Play more, and see that there appear more defects when the starting pattern has similarities to three (or more) learned patterns. In other words, the final recognized pattern may be significantly distorted if the initial pattern was not exactly in the appropriate *domain of attraction*. Conversely, if there are many learned patterns, their domains of attraction tend to overlap each other. We may compare this situation with a period of hard cramming for an exam—if we cram too much, then it is likely that we shall find ourselves confused during the test; almost certainly we shall reproduce some distorted pictures!

There are many more analogies between the dynamics of the Hopfield model and way our brains work. They all spring from the fact that **"parts" of the Hopfield model can be compared with the "parts" of human brain.**

5.6.2 The Hopfield model as a model of neural networks

Brains of living creatures consist of an enormous number of nerve cells, called **neurons**. In the human brain, there are about 100 thousand million neurons and each of them makes, on average, thousands of connections with other neurons through the "parts" called **synapses**.

A neuron emits electrochemical pulses of various frequencies, but in general these frequencies are either very high or very low. In a model of neural networks, we assume that a neuron possesses only two distinct states—a state in which it actively sends a signal and a state in which the neuron is inactive. The active state of a neuron, at a particular site i, can be identified with the spin oriented upward ($S_i = +1$), whereas the inactive state of the neuron can be identified with the downward oriented spin ($S_i = -1$).

A synapse may modify the signal being sent by the contiguous neurons. The amount of modification depends on the chemical properties of the neighboring environment. But there is strong evidence that properties of synapses can be modified by signals passing through them. Therefore the bond, or the coupling constant J_{ij}, of the Hopfield (or Ising) model can be identified with the strength or intensity of the synaptic connection between the neuron at site i and the neuron at site j. Accordingly, the learning mode of the neural network corresponds to training the network to store information (patterns) by inducing specific physical properties in the synaptic connections J_{ij}. Similarly, as you remember, in the Hopfield model the process of learning is modelled and effectuated by assigning values to the interaction parameters J_{ij} according to the rules described in Computer Activity 5.

COMPUTER ACTIVITY 7: PATTERN RECOGNITION — REAL PICTURES

[a] This activity demonstrates how the Hopfield model can be used for more sophisticated pattern–recognition tasks. The basic idea and framework is the same as in the preceding activities, except for the detail that your "playground" now appears in the left (light blue) part of the screen, while the learned patterns appears on the right. Besides, the energy landscape curves are not shown in this activity. The underlying spin system is now much larger (63×63 spins) than any used before in this workbook, and consequently the patterns may now look more like real pictures that one tries to recall. There are several preloaded pictures that you can use to investigate the system's properties (memory capabilities), but you may prefer to draw your own pictures. However, since it is hard to discern individual spins and click on particular ones (as they are many of them), in this case you can slowly drag the mouse (button pressed) across your "playground" and draw a picture, as with a pencil.

[b] Open the SPIN GLASS program and choose the option "Hopfield Model." Within the new screen image, click on the option "Hopfield

DEMO," and select **LoadPicture** from the menu **ShowPicture**. You
see the dialog box in the lower left portion of the screen, where you
can choose one of the preloaded picture files (their names end with
d, for *digital*). Choose, for example the file "jeepd" and click on the
option "Open." You see the chosen picture (pattern) displayed on
your "playground." Now, you can make the spin system (that is built
into the program) learn the chosen pattern (by adjusting its magnetic
bonds). To do this, select **Learn** from the **Options** menu. The system
then learns the chosen pattern and its image appears in the right third
of the screen.

Using the foregoing procedure, you can make the system learn another picture, say the one with the file name "girld". The new image
appears next to the first one learned, that is, in the right third of the
screen. Now we can say that the system has memorized two patterns,
and we may wish to test its memory. To this end, load one of these pictures again (but skip the learning stage) and use the **FlipRandomly**
command several times to distort the picture to the extent that it is
hard (for us!) to recognize its original image. Let's see whether the
"trained" spin system (the neural network model) is able to recognize
the distorted picture. From the **Options** menu choose the **Retrieve**
command several times, until the displayed image on your "playground"
does not change. Does the image correspond to the original, that is,
to one of the two memorized patterns? If it does, how can you explain
this admirable feature of the trained system? Of course, you are right
if you say that it is because the trained system possesses two energy
minima and it simply evolved to the proper one.

[c] Now, let us see whether the system distinguishes between the two
learned patterns. Is it possible that this "artificial brain" cannot tell
the difference between an image of a human face and an image of a
vehicle? Let's check! Load the third pattern (without going through
the learning stage). Load, for example, the pattern stored in the file
"card" (digital car), and use the **Retrieve** option until it no longer
brings any changes in the pattern displayed on your "playground." Do
you see the girl or the jeep?

What grade do you give the memory of the spin system as a model
of real neural networks? If you are reluctant to give it the grade "A",
test it again. Load the pattern stored in the file "maxd," and use the
"Retrieve" option as many times as it makes changes in the image that
appears on your "playground." What do you see now, the jeep or the
girl? Yes, Maxwell was the famous physicist and he was a human being
and his picture is closer to the picture of the girl than to the picture of

the jeep. We see that the system is able to distinguish between vehicles and human beings. But, be careful: the system, like a human being, has its limitations!

[d] Go ahead and make the system learn more than two patterns. When several patterns are stored in the system's "memory', load one of them again (without going through the learning stage) and distort its image by using the **FlipRandomly** option. Then try to regain the original image by using the **Retrieve** option. Can you always restore the initial pattern with no faults? Usually the more similar the learned patterns are, the harder it is for the system to retrieve the correct pattern from the distorted image. We can explain this by the fact that the learned patterns, being similar, correspond to energy minima (valleys in the energy landscape) that lie very close to each other and the system can get "confused" about which minimum it should enter. As a result, there appears (on the screen) a mixture of several learned patterns, a spurious image.

There are some more options in the **Options** menu, namely **Replace Pattern** (to indicate the pattern to be removed so that a new one can be stored in), **Unlearn Pattern** (to erase a pattern from memory), and **Get Picture** (to create a new picture by drawing it yourself). Play with all these options. In so doing, you are exploring near the boundary of what is known in the modern science and technology of *associative memory* and *pattern recognition*.

5.7 Bibliography

1. D. L. Alkon, "Memory Storage and Neural Systems", *Scientific American*, July 1989, p. 26.

2. P. J. Denning, "Neural Networks", *American Scientist*, September–October 1992, p. 426.

3. R. Menzel and J. Erber, "Learning and Memory in Bees", *Scientific American*, July 1978, p. 102.

4. D. L. Stein, "Spin Glasses", *Scientific American*, July 1989, p. 36.

5. D. W. Tank and J. J. Hopfield, "Collective Computation in Neuronlike Circuits", *Scientific American*, July 1987, p. 104.

6. J. A. Anderson and E. Rosenfeld, editors *"Neurocomputing: Foundations of Research"* (MIT Press, Cambridge, 1988).

7. J. Hertz, A. Krogh, and R.G. Palmer, *"Introduction to the Theory of Neural Computation"* (Addison–Wesley, Redwood City, 1991).

Contents

Chapter 6. Lightning and Soap Films

Chapter 6

Lightning and Soap Films

6.1 Introduction

We are sure that the title of this chapter sounds bewildering when you read it aloud. What has lightning to do with soap films? Lighting is something mighty, whereas soap films, like soap bubbles, are fragile and easy to blow away. So, is there a joke in the title? No, there is a well-established mathematical theory that constitutes the background for describing the lightning hairy shapes and shapes of soap films (or some rubber sheets) stretched over a firm metallic (or wood) frame. It is our goal to make you acquainted with this beautiful and versatile mathematical theory, and, in doing so, to build a model of fractal growth. To this end, let us start with the fractal that you can see in Figure 1.

Lightning is a form of *dielectric breakdown (or discharge)* that takes place in the atmosphere (which is a dielectric medium) and spans between base of a cloud (usually negatively charged) and some object (lightning rod, preferably) on the ground. In most cases lightning appears as a fractal shape (see Chapter II of this book), but it is difficult to analyze its properties because of the nonhomogeneity, humidity and electrical conductivity of air. The fractal shape from Figure 1 is an example of *dielectric breakdown* that is more accessible to study. It has been produced in a physical laboratory by exposing a block of plastic (Lucite) to a beam of accelerated electrons. The electrons have been stopped and trapped within a thin layer (about $1cm$ below the surface). Then came part of the experiment that you can **do yourself**—a nail was hammered in the bottom side of the plastic block, which caused electrons to shoot out, leaving the tree-like pattern. In fact, the 'branches', that is, the electron trails, are due to the heating (Joule) effect of the electric currents. So we come to:

Figure 6.1: A tree–like fractal structure produced by dielectric breakdown in a block of plastic.

QUESTION TO THINK ABOUT: Why did the nail 'collect' electrons, or more generally, why does a lightning rod 'attract' lightning?

The

answer to this question comes from the common knowledge of electrostatics, and it would go like this: The nail 'collected' electrons for the same reason that a lightning rod 'attracts' lightning, and the reason lies in the fact that the electric field is strongest at the tips (sharp points) of conducting objects. Now, you can insist and ask: What does it exactly mean that an electric *field* is strongest at certain places and how can one see it? We shall take the latter question seriously, and in answering it we shall find motivation to learn more about closely related problems in science and technology.

6.2   Electric Field

One of the very basic notions of modern physics is *field*. Simply speaking, a *field* is a part of space where certain physical quantities are defined at each point. For instance, you can speak about the temperature *field* in your room, and you have to admit that no matter whether you use a heating system (in winter), or an air–conditioner (in summer) you cannot achieve the same temperature at all points. Certainly, the temperature is always larger somewhere, and, at other points, it is lower.

6.2.1 Activity 1: Hands-on—Your Room

Take two sheets of paper and on both of them draw your room in the most simple way, as a cube for instance. Then, mark on one sheet, according to your own experience, hot, warm, cool, and cold places (regions) of your room during the summer time. On the other sheet do the same for your room in the winter time. If possible, use colorful pens— red (for hot places), orange (warm), yellow (tolerable temperature), green (cool), and finally, blue (cold). In this way, you will get pictures of the **temperature field** in your room. Of course, if you want to be precise, you can look for some sensitive thermometer and make daily observations inside your room. In the outside world, meteorologists do exactly the same thing by placing thermometers at certain fixed points on the ground (and in floating balloons). So, next time you watch a weather–forecast, pay attention to their pictures of the **temperature field** of *your state.*

Another example, instead of the temperature field, can be the pressure field in a swimming pool. Could you draw a swimming pool in a box–like shape, and mark layers of low and high water pressure? Thus, you will produce a picture of a different field, and it can be a very nice picture if you happen to use different shades of blue color for different layers of water in the swimming pool. But, let us turn our attention to the **electric field.** Where is it? What do you think?

The electric field appears in the neighborhood of every charged body. Each point in space around a charged body is characterized by a number Φ, called the potential, and by a vector $\vec{E}$, that is equal to the force acting on a unit positive point–charge if it is positioned at the given point. By convention, $\vec{E}$ is also called electric field. The potential Φ is equal to the work that one has to do (against $\vec{E}$) while bringing a unit positive charge from some reference point, usually taken to be far away (we say at infinity), to the given point.

It follows from the above bold lines that to know the electric field in the vicinity of a charged body, one should know the quantities Φ and $\vec{E}$ *at each point* of the surrounding space. This can be determined experimentally, by performing physical measurements, and theoretically, by doing proper calculations. Fortunately, Φ and $\vec{E}$ are closely related, and if we know one of them, then in principle, we know the other. Let us focus on Φ and set the goal to make pictures for various electric fields (of the type you have made for the temperature field in your room).

OUR MAIN TASK: We want to describe electric fields of various
charged bodies in terms of the potential Φ and to represent our
findings through pictures vividly depicting regions of different
values of Φ.

6.3   Electrostatics

Our *main task* belongs to the branch of science called *electrostatics*
which, literally speaking, deals with phenomena related to nonmoving
electric charges. One might say that it is a classical branch of science.
However, in recent years electrostatics has expanded its scope to such
applications as electrostatic precipitation of industrial wastes (such as
fly ash), electrostatic separation of mixed granular solids, electrostatic coating (painting), and—most interestingly—electrostatic imaging
(such as xerographic processing). We can rightly add (and demonstrate
later on) that knowledge of the laws of electrostatics has helped scientists make successful models of fractal growth. But, in all these cases
it was necessary to learn the potential Φ of the relevant electric fields.

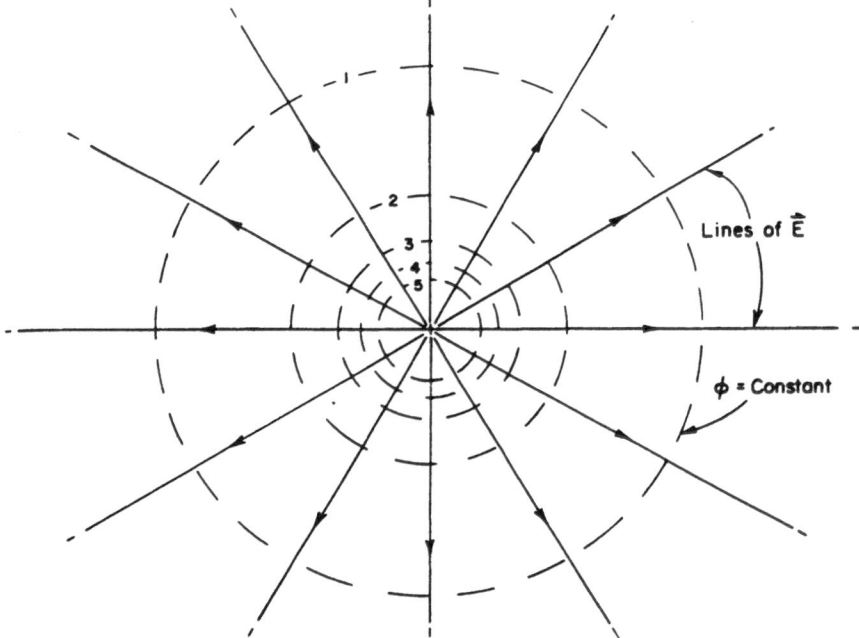

Figure 6.2: Geometrical representation of the electric field of a positive unit charge
(according to *The Feynman Lectures on Physics, Vol. II*, Addison-Wesley, Reading
MA, 1964).

In Figure 2 you can see a simple drawing that depicts the electric field of a positive point charge, that is, of a very small charged sphere whose radius we can neglect. Points of constant potential Φ constitute concentric spheres, but because we have made a cross–section through the field they appear as concentric circles. Hence, all points on a circle have equal potential, and, accordingly, we call them *equipotential lines*. Radial straight lines in this case represent the so–called *force (or field) lines* which can be interpreted in two ways: 1) At each point of a force line the vector $\vec{E}$ is tangent to it, and 2) A force line can be visualized as the path of a tiny positively charged particle (with almost zero mass) that travels through the space being constantly pushed by the field (the arrow on the line indicates direction of the particle moving).

QUESTION TO THINK ABOUT: Do you find drawing given in Figure 2 consistent with the Coulomb's law which states that force between two point charges q_1 and q_2 is directed along the line that connects them and that it is repelling if the charges have the same sign, and attractive if they have different signs? For your further thinking, we add that Coulomb also found that the force between two point charges is proportional to $1/r^2$, where r is distance between the charges.

In depicting electric fields scientists insist on the convention that the density of force lines in a vicinity of a given point represents the strength of the field at that point. More precisely, by the density of lines we mean the number of lines per unit area of a surface *perpendicular to the lines*. We *urge you to accept this convention*, and, as a simple exercise, to check whether it (the convention) has been correctly followed in Figure 2.

COMPUTER ACTIVITY 1: LIGHTNING ROD

[a] Move on to the computer program called CHILL OUT. Click on "File" and choose option "Open", and you will reach the folder named "Album".

[b] For the time being, skip other items and click on the "Lightning rod". The computer will load a drawing of a small solid circle being on the rod that is posted on a big open circle. Now, imagine that this system represents a charged body and you want to find the electric field around it.

[c] To get a picture of the field, in the menu item click on "Lattice", and in the submenu click on the item "Solve". After a few minutes you will get a colorful map displaying regions of different potentials, that is, you will get the potential field.

[d] Now, go again to the item "Lattice" and click on "Force lines". Soon you will see the picture of the type presented in Figure 3. In this figure, regions of different potential are not differently colored, but, are rather differently shaded in gray.

[e] If your picture is not exactly the same as the one in Figure 3 and you want to improve it, choose item "Attributes" in the "Lattice" menu and the dialog box will appear. Within the dialog box choose a larger number for the variable "Size" and a larger number in the denominator of the parameter "Epsilon", and go back to the item "Solve". Repeat the latter choosing until you get the picture that makes you satisfied.

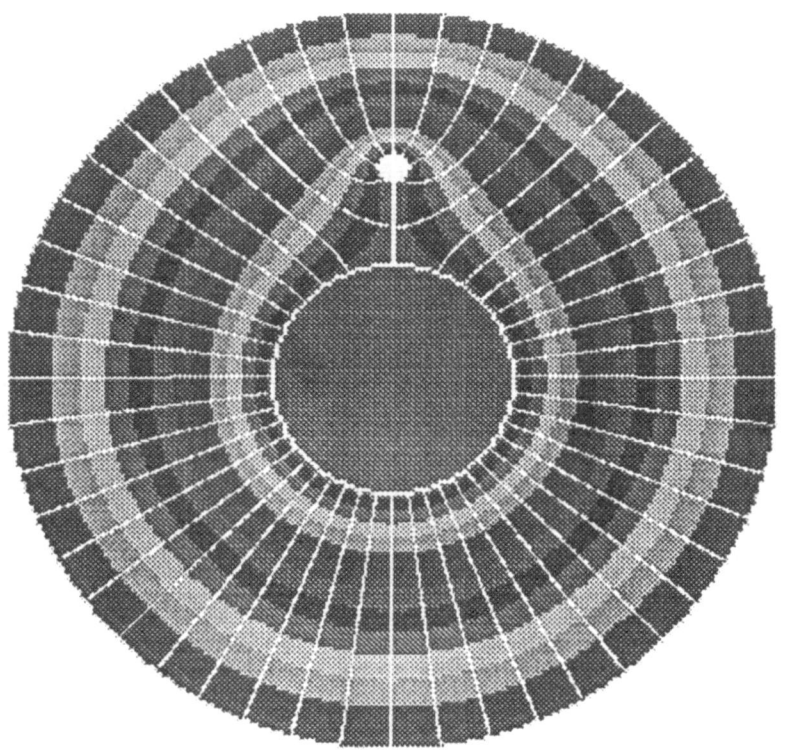

Figure 6.3: Electric field around a simple model of the system comprised of the earth (big circle) and the lighting rod (the post and the small ball on top of it).

LET US ANALYZE OUR RESULT: Later on we will give you many more opportunities to play with the program CHILL OUT in order to get familiar with it, and, in particular, we shall explain the way the program solved our problem. However, let us first analyze our joint finding.

In Figure 3 we have in fact a model of the lightning rod. Of course, everything is a little bit out of scale—the earth is too small compared to the rod, the ball on the top of the rod is not a sharp point, an so on. But, please do remember, this is a *model*, and, like a good caricature, it is good if it reveals the main features of the original. In the case under study, the model really does—if you remind yourself of the convention that the density of the force lines in a vicinity of a given point represents the strength of the field at that point, then you can see from Figure 3 that the electric field at the surface of the earth is the weakest in the region neighboring the post. Hence, the electric charge coming from the atmosphere (for instance, from the bottom of a charged cloud) will hit more likely any other place than the one close to the post. On the other hand, judging according to the number of force lines close to the top of the lightning rod, charges from the atmosphere will most likely hit the top of the rod. What will happen next? What do you think? If the rod is made of a conducting material, as it should be, the "collected" charges will be conducted to the ground, without any damage whatsoever. And, that is all. In short, it is how the lightning rod operates.

6.3.1 Activity 2: Hands-on — The Earth

Yes, we know that we have left you with some unanswered questions. First, the force lines in Figure 3 are not oriented (there are no arrows), and a few pages back we said that they should be. Now, we want to tell you that the earth can be conceived as a large conductor with a *negative* surface charge of about 450,000 *coulombs*. Assuming that there is no other charged body in the vicinity of the earth, add small arrows, pointed in the appropriate directions, to the force lines in Figure 3.

Have you noticed that the potential is the same everywhere on the earth surface and on the lightning rod connected to it? This is a very important property of conductors—electric potential is constant on their surfaces (as well as in their interiors).

Finally, here comes a kind of teaser. The earth in Figure 3 is depicted as an empty circle, meaning that one might imagine it as a hollow conductor. What do you think—what changes would happen in the picture of the electric field if we had depicted the earth as a solid conductor (filled circle)? To gain a firm answer to this question, you will find it helpful to repeat the preceding computer activity for two cases offered in the "Album": (a) for the "Empty circle," and (b) for the "Filled circle."

6.4   Laplace Equation

How does one learn the potential Φ of charged bodies of arbitrary shapes? This is the question that has been of the utmost importance in designing scientific instruments and in making new products, for instance, in the car and aircraft industry. In the case of bodies such as a point charge and a charged sphere, one can rely solely on the Coulomb law (nothing more was needed to make the picture of the type presented in Figure 2). In the case of more complicated bodies one needs more sophisticated mathematical tools, which spring from thoroughly elaborated statements of the old Coulomb law. Are we going to teach you those statements? No! We shall try to teach you their consequences, that is, those wonderful recipes of calculating Φ that you can understand and **apply** in the same way as a professional scientist (or an engineering practitioner) does.

THE RULE OF THUMB IN SCIENCE: To solve a small or a big problem, you have to formulate it carefully in the most precise way that is possible.

Following this rule, we dare repeat that we want to know the potential Φ at *every* point in a part of the space adjacent to charged bodies. How do we specify a point?

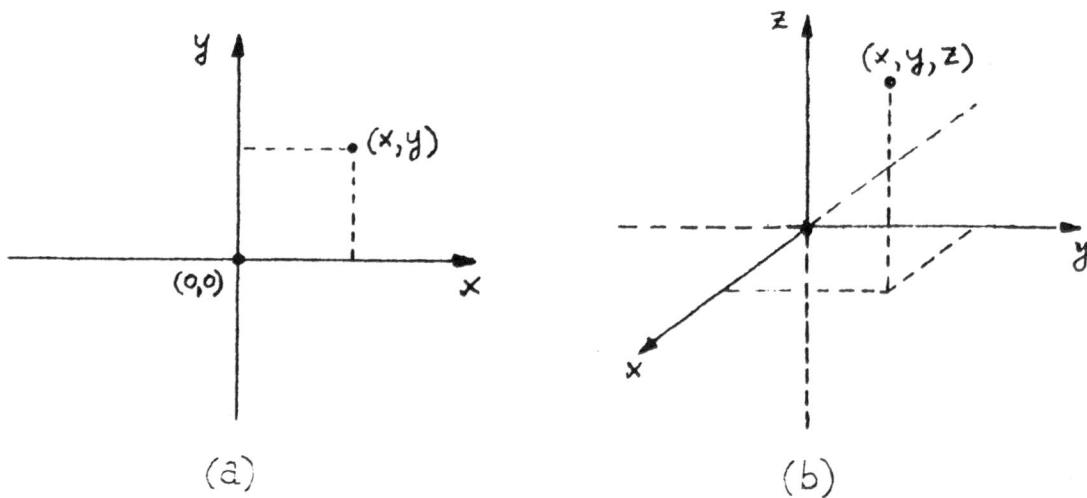

Figure 6.4: A point in the two–dimensional (a) and three-dimensional (b) space.

The location of any point in a space can be specified by choosing first some particular point as the *reference point* (something like a downtown of a big city). Next, we introduce the so–called *coordinate system*, which in the case of planes (the two-dimensional spaces) is a simple system of two perpendicular lines whose crossing (*coordinate center*) should be placed at the reference point and marked by (0,0), that is, by a set of two zeros (see Figure 4a). Any other point in the two-dimensional space is also specified by two numbers (x, y), such that x measures distance right (positive x), or left (negative x), to the center, whereas y measures upward ($y > 0$), or downward ($y < 0$), distance of the given point from the center. Similarly, in the three-dimensional space we introduce a coordinate system that consists of three mutually perpendicular lines, whose crossing is the center $(0, 0, 0)$ and any other point is specified by the set of three numbers (x, y, z) (see Figure 4). For any particular point (x, y, z), value of z can be visualized as vertical distance of the given point from the (x, y) plane.

Let us go back to our main task. In practice, we can be challenged to study, in some cases, the electric field in three-dimensional space, and in the other cases we may need to learn properties of an electric field in two-dimensional space. Then, in the former case, we can say that we know the field if we know the value of Φ at each point (x, y, z), while in the latter case we should know Φ at each point given by (x, y). In order not to use many words, mathematicians simply say that Φ can be function either of all three coordinates x, y, and z (in the three-dimensional case), or a function of x and y (in the two–dimensional case), by writing it as $\Phi(x, y, z)$ and $\Phi(x, y)$, respectively. Therefore, we now can give now a more precise formulation of our main task stated on page 3—**we want to calculate the function Φ for various electric fields**.

QUESTION TO BE ANSWERED PROMPTLY: Have we formulated our main task in the most precise way? Have we followed the Rule of Thumb given on the previous page?

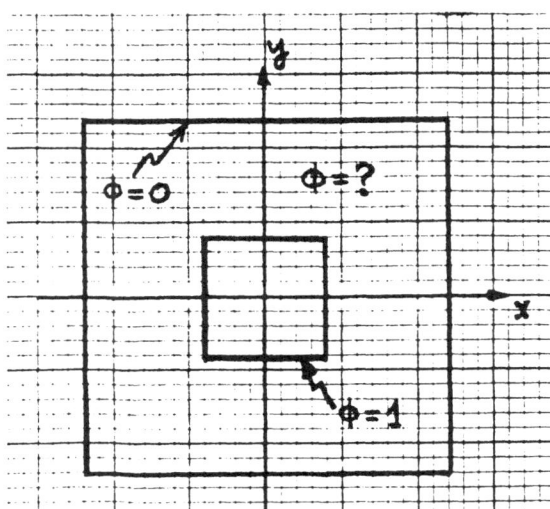

Figure 6.5: What is Φ between the two differently charged square frames?

We must admit that we have not yet formulated our main problem quite precisely. Why? What else do we need to say? We need to say in what region of space we want to calculate the function Φ. Most usually, it is taken to be the region two differently charged conductors. Or, since the potential Φ is the same on the surface of a conductor (as you know from the `Activity 2`), we can say that we want to calculate Φ in the region between two equipotential surfaces. It could be, for instance, a region between two spheres, or, more simply, region between square frames made of a wire, that are differently charged and thus have different potentials (see Figure 5). In this way, our problem is precisely formulated—we want to calculate Φ in the region between two equipotential surfaces (or curves) where there is no other charge.

A SOPHISTICATED TOOL: Our precisely-defined problem means that one has to solve the following equation

$$\frac{\partial^2 \Phi}{\partial x^2} + \frac{\partial^2 \Phi}{\partial y^2} + \frac{\partial^2 \Phi}{\partial z^2} = 0, \qquad (6.1)$$

for the problem in three–dimensional space, or the "simpler" equation

$$\frac{\partial^2 \Phi}{\partial x^2} + \frac{\partial^2 \Phi}{\partial y^2} = 0, \qquad (6.2)$$

in the case of the problem defined in a two-dimensional space. An equation of this type is called the Laplacian equation, according to the French scientist (mathematician, physicist and astronomer) P. S. de Laplace (1749–1827).

Although we are not going to teach you how to 'derive' the Laplace equation, nor how to solve it, we have put it here for three reasons: (i) to acquaint you with its appearance, so that one day when you start to major in science or engineering you can regard it as an old friend; (ii) when you hear scientists speak of it while considering problems of the type we have called "our main task," you will know what they are talking about; and, most importantly, (iii) because it has been shown (through the joint efforts of many scientists that the solution of "our main task" *is equivalent* to the solutions to a variety of different physical and technical problems, we will demonstrate and exploit this equivalence in a simple way—but you should know that the most exact way of proving it is by invoking the Laplace equation.

What do we mean saying that the solution of our main task can be *equivalent* to the solution of some other problem? Let us offer you one example from a realm that have attracted many great minds and talents in the past, including almost every child, and most likely it will be the same throughout the future. What is the realm? It is the realm of soap bubbles and soap films. To make a long story short, we urge you to try to make two square frames of wire and connect them in the way it is

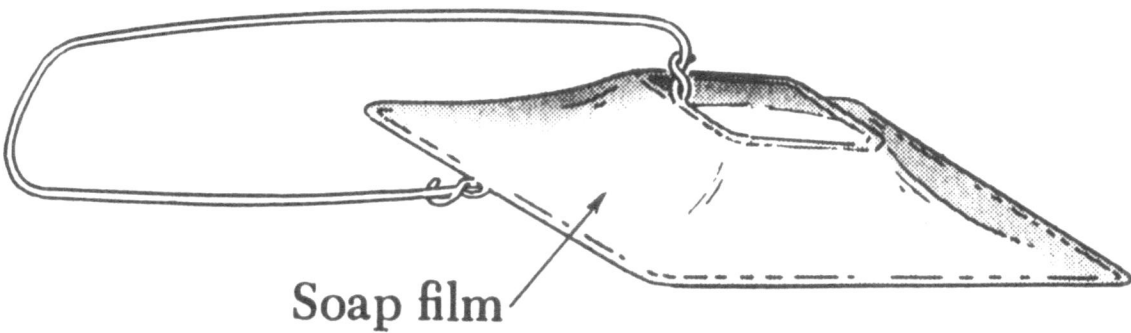

Soap film

Figure 6.6: The soap film that provides an analogous solution of the electric problem depicted in Figure 5 (according to E. M. Purcell *"Electricity and Magnetism"*, Mc Graw–Hill, New York, 1965).

shown in Figure 6, so that the smaller square is 1 *inch* above the larger square. Then soak the two square in a soapy liquid—the soap film that will span between the two squares will provide **an analogous solution** of the electric field problem depicted in Figure 5. How? Why? What is analogous? If you place the film above the graph paper, then the height h of the soap film above each point (x, y) **is equal to** the value $\Phi(x, y)$ of the potential. Why is it so? The *sophisticated answer* would be—it is so because both h and Φ satisfy the same Laplace equation (2) with the same boundary conditions, which means that the difference of Φ, as well as of h, between the inner and outer frame was set to be equal to one *unit*. A more appealing answer to the puzzle of equivalence would be—values of the potential Φ at each point (x, y) should reflect the *equilibrium* distribution of electric charges on the conducting frames. On the other hand, values of the height h of the soap film at each point (x, y) should reflect *equilibrium* of elastic forces within the film.

6.4.1 Activity 3: Hands-On: An Old Recipe

It is hard to believe, but we have found a very good recipe for making soapy liquid, for producing bubbles and soap films, in the book *"What is Mathematics?"*, written 50 years ago by two eminent mathematicians— R. Courant and H. Robbins. Here is their recipe: *"Dissolve 10 grams of pure dry sodium oleate in 500 grams of distilled water, and mix 15 cubic units of the solution with 11 cubic units of glycerin. Films obtained with this solution and with frames of brass wire are relatively stable. The frames should not exceed five or six inches of diameter".*

We invite you to follow the recipe and make the soapy solution (and the frames shown in Figure 6!). If you cannot find pure sodium oleate, you may use some detergent powder. However, it is important to add glycerin in the way it is explained in the recipe. Finally, we would like to ask you something—how would you explain the fact that the recipe of this kind appeared in a book on mathematics? As a hint, we can mention that the recipe is given in the section entitled *"Experimental Solutions of Minimum Problems".*

There are other interesting problems whose solutions appear to **be equivalent** to the problem of finding the potential function Φ. A very inspiring one is **the random walk problem** you learnt about in the first chapter of this book. We are going to illustrate the equivalence in the case of the simple Φ problem shown in Figure 5. Imagine a random walker that walks between the two square frames—it could be a drunken sailor on a floating platform, such that the small solid square represents the place with cabins and beyond the large square there is water everywhere. The sailor starts at some point (x, y) and you ask what is the probability $p(x, y)$ that he will reach the cabin before he falls into the water. Of course, we assume that the drunken sailor can make a step with equal probability $1/4$, from each point where he happens to be, to four neighboring points (left, right, forward, and backward). It turns out that **the function** $p(x, y)$ is equal to the function $\Phi(x, y)$, and, for the reasons explained previously, equal to the function $h(x, y)$ in the case of the soap analogy.

A BIT OF INSPECTION: If $p(x, y) = h(x, y)$ and $h(x, y)$ represents the height of the soap film shown in Figure 6, then we can see that the probability that the sailor will reach safely his cabin is larger when he is closer to it (the smaller solid circle in Figure 5). On the other hand, h is smaller close to the outer frame meaning that the sailor is more likely to fall into the water when he starts staggering at the verge of the platform. Does it make sense to you? In particular, what do you say about $p(x, y) = 0$ on the platform edge, and $p(x, y) = 1$ at the entrance to the cabin?

Our final example of the problem of equivalence is from thermal physics—the potential Φ problem has the same solution as the problem of finding temperature T at each space point, providing that the inner and outer boundary shapes and values are the same. In this case, it means that we assume that between the boundaries there is a homogeneous medium, and two constant temperatures at the boundaries. For instance, to make correspondence with the example given in Figure 5, we can assume a thin metal plate having a square cut out from the center (see Figure 7), with temperatures $T = 0$ and $T = 1$ kept con-

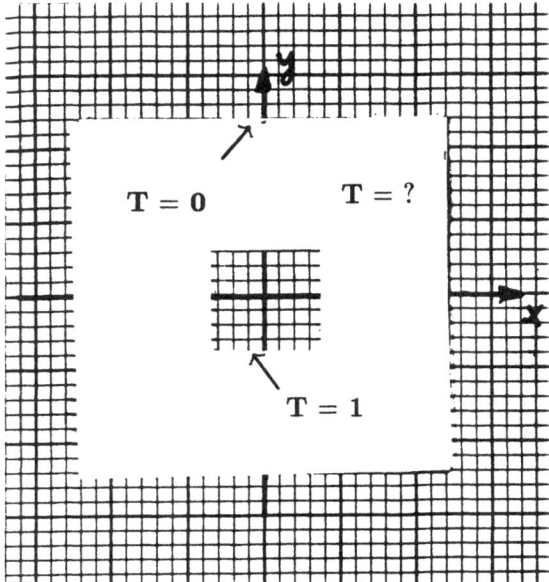

Figure 6.7: What is the temperature $T(x, y)$ at a point (x, y) of the metal plate whose inner and outer edges are kept at the two constant temperatures?

stant at the outer and inner edge of the plate, respectively. Thus, at each point (x, y) the equality $\Phi(x, y) = T(x, y)$ is valid for the problems defined in Figures 5 and 7. Why is it so? How can electric potential be equal to temperature? Be careful! More precisely, we should say that the values are equal, but the units, of course, are quite different. But, still, why are the values equal? Again, *the sophisticated answer* would be that it is so because both h and Φ satisfy the same Laplace equation (2) with the same boundary conditions. On the other hand, *the more appealing answer* would invoke the fact that in the case of the heated metal plate we are dealing with a special kind of equilibrium—*a steady state* of heat transfer (from inner edge to the outer edge of the plate). At this point we would not be surprised to see you stand up and ask: *"Is there something in between the sophisticated and the more appealing answer? Is there something that we can do ourselves and make us convinced that what has been said is right?"*. Yes, there is. Here it comes. However, before we start the new section, let us emphasize that we shall focus our attention in the rest of this chapter (and in the entire program CHILL OUT) on the two-dimensional fields.

6.5 Numerical Approach to our Main Task

All the quantities we spoke about in the preceding section, that is, the potential function $\Phi(x,y)$, the soap–film height $h(x,y)$, the probability $p(x,y)$, and the temperature $T(x,y)$, have a remarkable property that we are going to call **the mean–value property**, or simply the **MV property**. The property consists in the fact that the value of any of these functions at some point (x,y) is equal to the average of all the values that the function takes on a small circle whose center is placed at the given point. Functions that satisfy the MV property are called **harmonic functions**. If we use a fine grid, represented, for instance, by graph paper, to define points in the region of space of interest, then a point (x,y) can have only four neighboring points *at the same distance* a, where a is the so-called **lattice constant** of the grid (in the case of the graph paper a is usually equal to $1\ mm$). This allows us to give the simple mathematical expression for the mean–value property of the potential Φ, for instance,

$$\Phi(x,y) = \frac{1}{4}[\Phi(x+a,y) + \Phi(x-a,y) + \Phi(x,y+a) + \Phi(x,y-a)]. \quad (6.3)$$

Of course equation (3) is merely an approximation because we are taking only four out of multitude of points that could lie (if there were no grid=lattice) on the distance a from the given point. Yet, it is a very powerful approximation, and hence, in the case of the discrete description of Φ (and similar functions) we get the simple formulation of the MV property: **The value of the function at a point is equal to the arithmetic mean of the values that the function has at the four neighboring points**. Needless to say, the MV property springs from the Laplace equation, but it can also be associated with the fact that all the functions in the question describe (at each point) either a certain type of local equilibrium or a steady state property.

6.5.1 Activity 4: Hands-On: Field Research I

In Figure 8 we provide for you the lattice type representation of the potential function Φ, but it can be any other function we mentioned earlier, providing that it satisfies the given boundary conditions (which are the same as those in Figure 5, except for the fact that here we take that the small square is extremely small). In the case of the electrostatics, we can interpret the picture as a field of the Φ values that correspond to a point-like positive charge surrounded by a metal frame which is kept at the zero potential. In the case of the drunken sailor, we can visualize this picture as a set of probabilities p such that the value at each point represents the likelihood that the sailor will

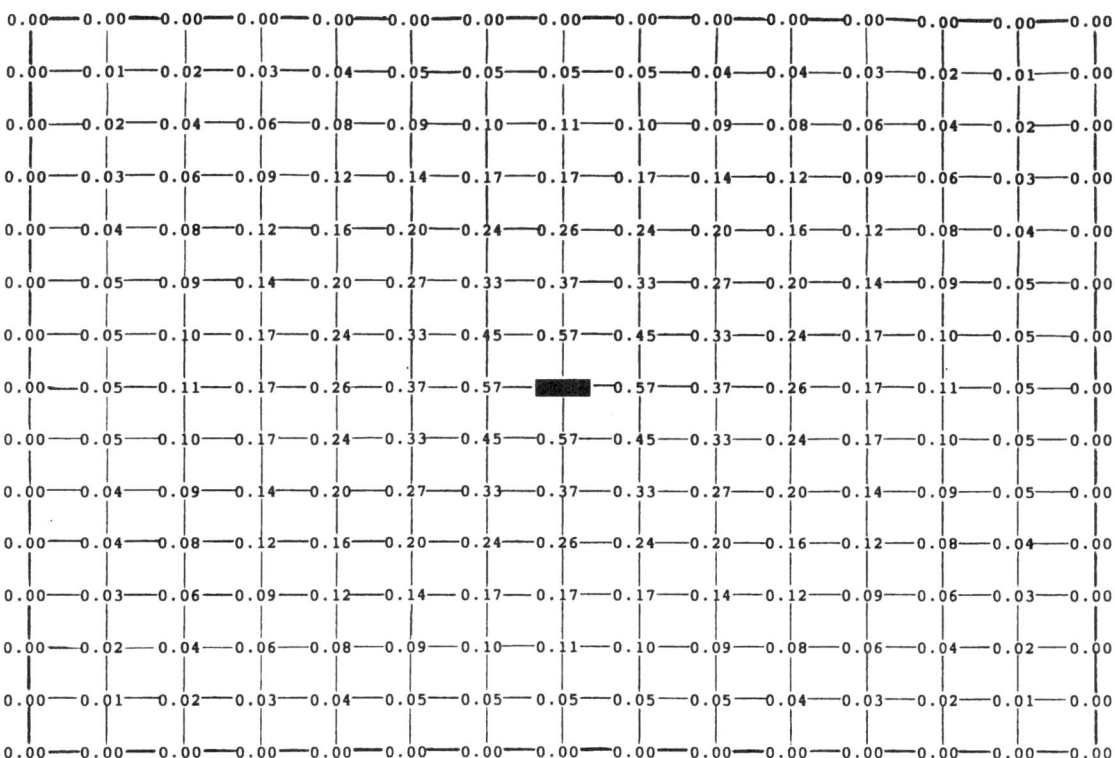

Figure 6.8: The lattice type (numerical) representation of the potential Φ of a positive point-like charge in the center of a metal frame that is kept at zero potential. The point-like charge is represented by the small solid bar, while Φ is taken to be equal to 1 at the point $(x = 0, y = 0)$.

reach his cabin (at the center) and will not fall in the water (at the rim).

Check, at as many points as you can, that the MV property, given by Eq. (6.3), is satisfied. Start, for instance, with the value 0.57 at right to the center and check that the sum $(0.37 + 1 + 0.45 + 0.45)$ divided by four is approximately equal to 0.57. Or, start with the lower right corner, at the value 0.01, and you can immediately see that when you divide $0.02 + 0.02$ by four you get 0.01.

You can also draw the equipotential lines—take several pens (different colors) and mark the same numbers around the center with one color. For instance, color all points with the 0.57 value in red, then use a yellow pen to mark all points with 0.45, and so on. When you exhaust your set of colors, start with the red again, that is, use your set of pens in cycles. When you are finished with marking the same values, go back and connect them with continuous broad curves. In drawing the curves—the equipotential lines—you should take care that, for instance, a curve which joins two point with the 0.45 values should pass between 0.37 and 0.57, and not between 0.57 and 1.0 (the center). If you get finished with the whole picture in one run, we congratulate you on mastering it. However, your satisfaction with producing a kind of picture that usually requires computer work will be its own reward.

QUESTION YOU SHOULD ASK: How does one calculate the values given in Figure 8? Can we learn to do it? Is it difficult?

The method that makes it possible to calculate a harmonic function, Φ for instance, is called the **relaxation method**. It is not difficult, and you can certainly master it. More than that—you will like it! In fact, the relaxation method is based entirely on the MV property of the harmonic functions

Let us start with the simple example that is given below, and which is quite similar to the large lattice of Figure 8—the only difference is that the boundary conditions are reversed: at the outer edge Φ is equal to 1, whereas in the center it is equal to zero.

1.0	1.0	1.0	1.0	1.0
1.0	??	??	??	1.0
1.0	??	0.0	??	1.0
1.0	??	??	??	1.0
1.0	1.0	1.0	1.0	1.0

To learn Φ in the region between the 'boundaries', that is, at the points marked by '??', you can, at the beginning, set these points to **any value you may think of between the two boundary values**, and start to apply the MV property in a consecutive way. By applying the MV property, we mean to change each entry (except for the boundary ones) to the corresponding arithmetic mean according to the equation (3). Of course, it is better if you can make a good guess at the beginning since in that case you will reach a proper answer sooner. So, let us simply put 0.5 in place of all the '??', and consequently you will have to start your calculation with the following setup

You know that the 0.5 values are not correct and, in accord with the relaxation method, you should start to improve them by applying the MV property. You simply 'sweep' over the lattice, for instance, in the clockwise direction starting with the upper left 0.5 entry. You may use your pocket calculator, and after the first run you will get

1.0	1.0	1.0	1.0	1.0
1.0	0.5	0.5	0.5	1.0
1.0	0.5	0.0	0.5	1.0
1.0	0.5	0.5	0.5	1.0
1.0	1.0	1.0	1.0	1.0

1.0	1.0	1.0	1.0	1.0
1.0	0.75	0.56	0.76	1.0
1.0	0.63	0.0	0.56	1.0
1.0	0.77	0.57	0.77	1.0
1.0	1.0	1.0	1.0	1.0

This is a better 'solution', but still it is not a satisfactory one. For instance, 0.75 is not one fourth of the sum $(1.0 + 1.0 + 0.56 + 0.63)$. Therefore, you should do several more sweeps until your solution gets relaxed to the following

1.0	1.0	1.0	1.0	1.0
1.0	0.83	0.67	0.83	1.0
1.0	0.67	0.0	0.67	1.0
1.0	0.83	0.67	0.83	1.0
1.0	1.0	1.0	1.0	1.0

We can expect that there are those of you who could say that *"the solution is not completely relaxed"*. Yes, they can go on improving and learn more digits at each place (that is, at each point of the field), and in doing so they will learn that instead of 0.83 there should stay 0.83333 and instead of 0.67 there should stay 0.66666. In other words, they can improve on the final solution, which is, in the program CHILL OUT, achieved by choosing smaller values for the parameter "Epsilon" in the "Attributes" dialog box. Thus you have learnt how the relaxation method works, which is in fact the way the program CHILL OUT 'solves' the Φ problem.

QUESTION TO THINK ABOUT: What is the most advantageous feature of the relaxation method? What impresses you most about what it can do?

To our minds, the nicest feature of the relaxation method appears to be the fact that you can start from whatever initial guess, for unknown values of Φ, you choose, and yet, in the end, your solution will **relax** to the same final result (in mathematics, this fact is known as the Lindenbaum theorem). You can check this feature in the above example by choosing uniformly assigned value 0.4 instead of 0.5, for the initial setup. In addition, you can check the same feature by starting with *a set of random numbers*, say, a set of eight random numbers between zero and one. Or, you can proceed to the program CHILL OUT and choose various "Initialization" possibilities in the "Attributes" dialog box, and thereby get the computer to work for you.

COMPUTER ACTIVITY 2: COMPLETE ALBUM

[a] Now that you have learnt one of the most powerful ways to solve *our main task*, that is, the relaxation method for calculating potential Φ, you can move to the computer program CHILL OUT and go through the *complete album* of already drawn charged bodies (frames). You can also draw a body of any shape you find interesting. The advantage of going through the almost complete album is that in the end you will attain an educated feeling about the potential Φ (or, equivalently, about the height h, the probability p, and the temperature T) under various circumstances, and this feeling will help you to understand many phenomena that occur in nature and in a laboratory.

[b] We strongly suggest that you start with the example presented, in different ways, in Figures 5, 6, ,7 and 8, which means that you should click either on the "Empty square" or on the "Filled square", within the "Album" options. More importantly, within the "Attributes" dialog box you should choose the square border for the outer boundary shape. Thus you should get a picture of the type presented in Figure 9.

[c] Imagine that the picture in Figure 6 represents terrain with the square flat region on the top, and that 'flying over the terrain' you are assigned to depict regions with different altitudes below the top. Then, if you were a good cartographer (which you can easily be!) you would produce a picture of the type given in Figure 9. Of course, a cartographer would use different tones of brown and green color to represent regions with different altitudes. In our case, different altitudes can mean different potentials, different soap–film heights, different temperatures, and, most interestingly, different probabilities.

[d] Related to the previous teasing task, we cannot resist asking one more teasing question—imagine that you are assigned, as cartographer, to draw *lines of the steepest descend* from the top square rim. Why is it important, first of all? Let us say that in mountains such lines show ways a rolling stone, or an avalanche, can go down. So, what would you do? By now, you are experienced and you can figure out that what the gravitational field does in the mountains similar feats are produced by the electric field between charged bodies. Therefore, you will simply 'ask' the computer, that is, the program CHILL OUT, to draw "Force lines" in the case of Figure 9, and you will get the required

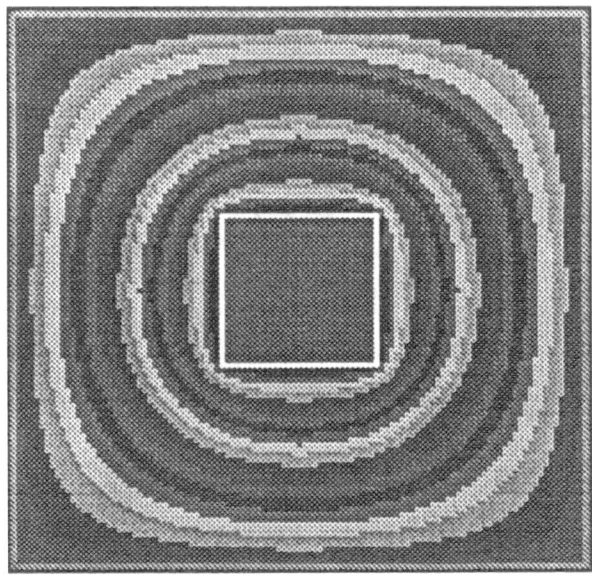

Figure 6.9: Solution of the problem defined in Figure 5, and in Figure 7 as well. The strips around the small square correspond to the regions with different values of potential (temperature, soap-film height, probability; see the text just before and after Activity 3).

steepest descent lines. Do observe that they are always perpendicular to the equipotential lines (see Figure 10). Finally, if you are tired take a break, but come back and check that a test particle will 'climb the hill', by clicking on the "Launch Particle" command. This means that it is a negatively charged particle because only such particles can go from a lower potential (large square) to a higher potential (smaller square).

6.5.2 Activity 5: Hands-On: Wood Work

[a] Try to find old wooden drawers, preferably one deeper than the other (we have seen many of them thrown in the streets!). Reshape them so that both of them get a square-like bottom, but try to make the deeper one have three times smaller side width than the shallow one. So, when you place a 'new' deep wooden box inside of the 'new' shallow box and look from above their rims should look like the squares in Figure 5. Now, take a rubber sheet (if you cannot find it, you may use a piece of a big elastic glove) and stretch the sheet over the rims of the boxes. Next, you should staple the edges of the rubber sheet to the rim of the larger box, by using four wooden slats (see the cross section of the set–up in Figure 11). What have you got? You have made **an analogous machine** for solving the problem depicted in Figure 5.

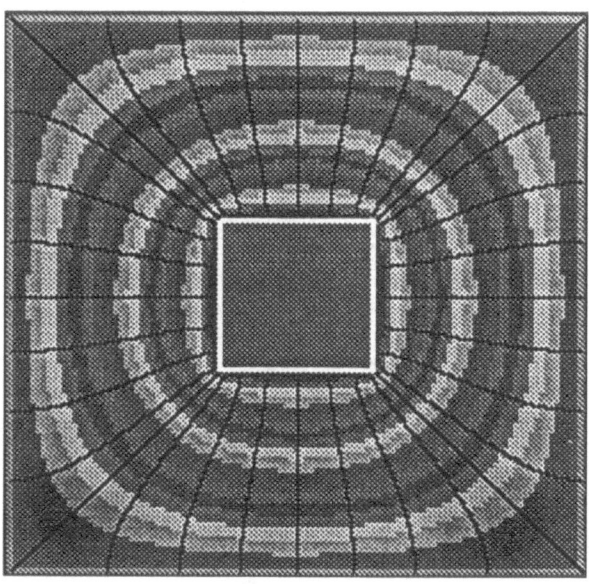

Figure 6.10: Force lines (black solid lines) of the field depicted in Figure 9.

Indeed, it is as much of a 'machine' as was the soap-film in Figure 6. Both 'machines' offer equivalent solution to the problem—the function Φ is equal to the height h of the rubber sheet (or to the height of soap film). The only difference comes from the fact a rubber sheet is less fragile than the soap film.

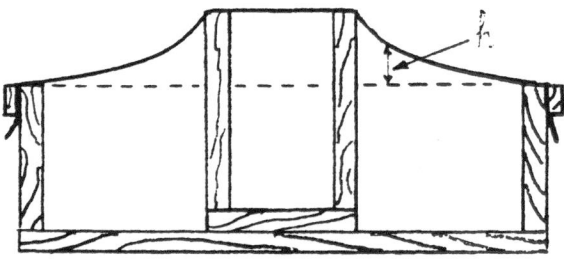

Figure 6.11: Cross section of the stretched rubber sheet used as an analogous machine for solving problem depicted in Figure 5.

[b] Once, you have made your machine, you can do experiments. First, you can get a physical answer as to why the numerical method we spoke about earlier was named the *relaxation method*. You can pinch the rubber sheet at various places (even many places at the same time), or you can push various rods and bars against the sheet, but once you let the sheet alone it will **relax** to the equilibrium position defined by

the boundary conditions (stapled slats). In other words, your initial conditions (the amount of pinching and pushing up or down) did not change the final shape of the rubber sheet, which is quite analogous to the fact that we referred to as the Lindenbaum theorem.

[c] Finally, you can take some light marbles (or some light plastic beads) and let them roll down the rubber sheet. Their paths should be in accord with the force lines found by computer and given in Figure 10.

In practice, for scientific and technical purposes, it used to be easier to make wooden models ('analogous machines') with the stretched rubber sheets, or metallic frames with stretched soap films, than to calculate the needed function Φ. Related to this approach, we quote here words of the late Nobel–prize winner Richard Feynman, that can be found in Feynman's *"Lectures on Physics"* (Volume II, Section *Electrostatic Analogs*): *"The stretched rubber sheet has often been used as a way of solving electrical problems experimentally...This method was used to design the complicated geometry of many photomultiplier tubes (such as the ones used for scintillation counters, and the one used for controlling the headlight beams on Cadillacs). The method is still used, but the accuracy is limited. For the most accurate work, it is better to determine the fields by numerical methods, using the large electronic computing machines"*. These lectures were delivered and printed in the early sixties. We have underlined the words *'complicated geometry'* and *'large electronic computing machines'* in order to demonstrate to you the progress of science that occurred in the intervening period and in which you are taking a part right now. Indeed, using the program CHILL OUT you can calculate and see the field of a charged object with a *very complicated geometry* (choose "Cluster 1" and "Cluster 2" from the "Album") **on a personal computer** and not on *'large computing machines'*. Moreover, using the same program and the same PC facility you will able to learn in the next section how *objects of complicated forms grow* in nature and in laboratories.

6.6 Fractal Growth

In this section we invite you to learn a model of fractal growth. All you need to conquer the model is your good will and decent knowledge of the material from preceding sections. Before starting to explain the details, we want also to remind (and warn) you that a model, like a caricature, is good if it reveals the main features of the original, and if it can be used over and over again, like an interesting and robust toy.

In 1984 a group of scientists from the research center of the Swiss industrial corporation *Brown Boveri* proposed a model to describe growth of patterns that appear during a dielectric breakdown (for your information, dielectric breakdown is just a kind of electrical-discharge in a nonconducting material). A typical pattern of the type that the Swiss group wanted to describe is given in Figure 12, and is known as the **Lichtenberg figure**, after the 18-*th* century German physicist Christoph Lichtenberg. You will not be wrong if you think of this figure, for the sake of simplicity, as a photograph of lightning taken during a thunderstorm. Concerning the irregularity of numerous white branches that you can see in Figure 12, we can tell you that an analysis (of the type described in the second chapter of this book) reveals that the branches comprise a fractal with the fractal dimension D close to 1.7.

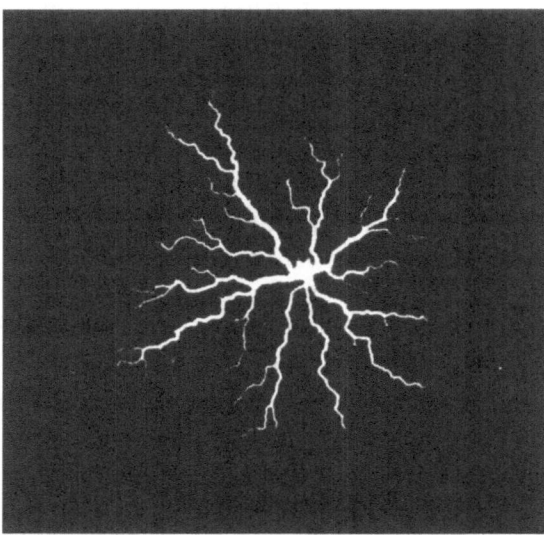

Figure 6.12: The *fractal dielectric breakdown pattern* is known in the history of physics as a Lichtenberg figure. It can be produced in a laboratory by applying a voltage to an electrode that touches a piece of a photographic emulsion on a glass plate, while the other side of the plate is covered by a grounded conducting material. The entire system should be surrounded by a compressed gas SF_6, and a high voltage ($30kV$) should be applied as a pulse (during a period that is no longer than a millionth of a second). Thus one gets the stringy electrical–discharge pattern.

The Swiss group invented a computer model that simulates the growth of patterns which are quite similar to the dielectric breakdown pattern of Figure 12. For this reason, the Swiss model is known as the Dielectric Breakdown Model, or simply the **DBM model**. We may say that the DBM model is merely a persistent stepwise application of the *relaxation method* you have mastered in the foregoing section. In Figure 13 the computer pattern is depicted after the 28-*th* step of calculation. At the beginning of the calculation there was just the central electrode at the zero potential ($\Phi = 0$), while the other electrode is modeled by the circular frame with $\Phi = 1$ (*Of course, the voltage difference being equal to one does not seem to be big enough, but don't forget that we are not using some specific units for the potential, and, most importantly, don't forget that we are dealing with a model*). Next, it is assumed that there is a finite (but not a small) number of available sites in the region between the two electrodes and that these sites lie on a square lattice. When the electrical discharge starts, the four nearest neighbors of the central electrode will be occupied with a probability 1/4, which *simulates the first small damage in the emulsion*.

QUESTION TO THINK ABOUT: Having two sites, the central and one of its four nearest neighbors, hit by the electrical discharge what would be the next site occupied by the discharge pattern?

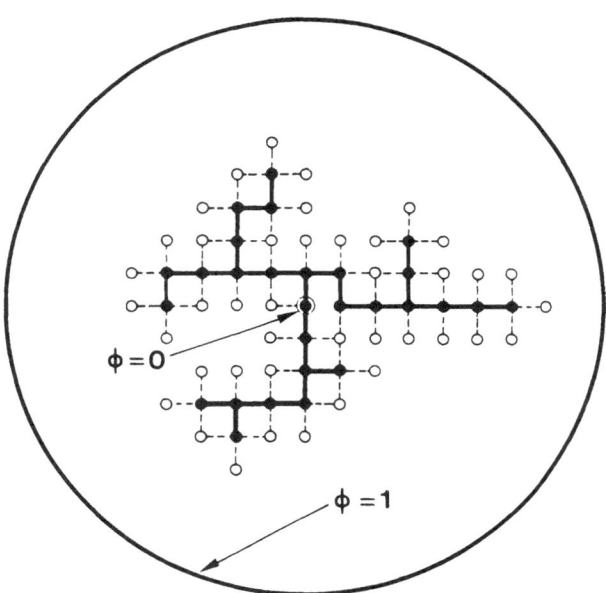

Figure 6.13: Details of the computer simulation of the fractal growth. The fractal discharge pattern is depicted by the 28 black dots and the solid black bonds. It is assumed that all of them are at the potential $\Phi = 0$, whereas the outer rim is kept at $\Phi = 1$. The next (the 29-*th*) point to be occupied is one of the nearest neighboring open small circles connected with dashed bonds to the existing pattern. The probability of 'capturing' one of the 41 open circles into the discharge pattern is proportional to the strength of the electric field. So, what do you think, which of the 41 open circles is most likely to become black in the next growth step?

You are right if you think that the location (x, y) of the next occupied site depends on the strength of the electric field. The Swiss group had the same thoughts, and in fact they assumed that the next occupied site would *most likely* occur where the electric field is strongest. So, in the next step they applied the *relaxation method* and calculated $\Phi(x, y)$ for all points, except for the two at the center (where $\Phi = 0$, by assumption) and at the circular frame (where $\Phi = 1$). Now, you can inquire—knowing Φ how do we calculate the strength of the electric field? Without making a long explanation, we may say that the strength of the field is given by the rate of change in potential at various points (observe, for instance, in Figure 3 the fast change, over a unit distance, of the equipotential lines close to the lightning rod, where the field is the strongest). Then, it follows that in the case under study the field at the points neighboring the existing pattern *is equal to the potential* $\Phi(x, y)$ at these points. **How? Why?** Simply, if we denote the location of points of the pattern by (x_0, y_0) than we know that $\Phi(x_0, y_0) = 0$, and hence the rate of change $\Phi(x, y) - \Phi(x_0, y_0)$ turns to be equal to $\Phi(x, y)$. So, the Swiss group, guided by the assumption that the probability is proportional to the field strength, suggested to take the quantity

$$c(\Phi(x, y))^\eta, \tag{6.4}$$

probability that the site at the location (x, y) will be next to grow. Don't get scared of the formula—the Greek letter η (eta) was introduced to make the model *the toy*) more versatile, and for the beginning you can take $\eta = 1$. On the other hand, the constant c stands for the necessity that all probabilities should add to one, which means that one of the neighboring site must with the certainty join the pattern. To satisfy the latter requirement, c should be taken equal to one divided by the sum of all $(\Phi(x, y))^\eta$ with (x, y) running over locations of the neighboring sites. **With this the DBM model is completed**—the next site to be occupied is chosen according to the accepted probability law. Therefore, *let us be cautious* and say that, in the case of only two sites occupied, the third site would most likely be captured so as to form a three–point line together with the previous two. In other words, the third site has good chances to appear at one of the two ends of the two–site cluster. Why? Because, as you know by now, the field is strongest at the tips (at the pointed ends). Similarly, in the case of the 28–site cluster of Figure 13, the 29-*th* site will most likely be captured into the pattern at one of the tips of its branches.

QUESTION TO THINK ABOUT: Why did we call for cautiousness in predicting the location of the next site to be occupied when we know that the field is strongest at tips?

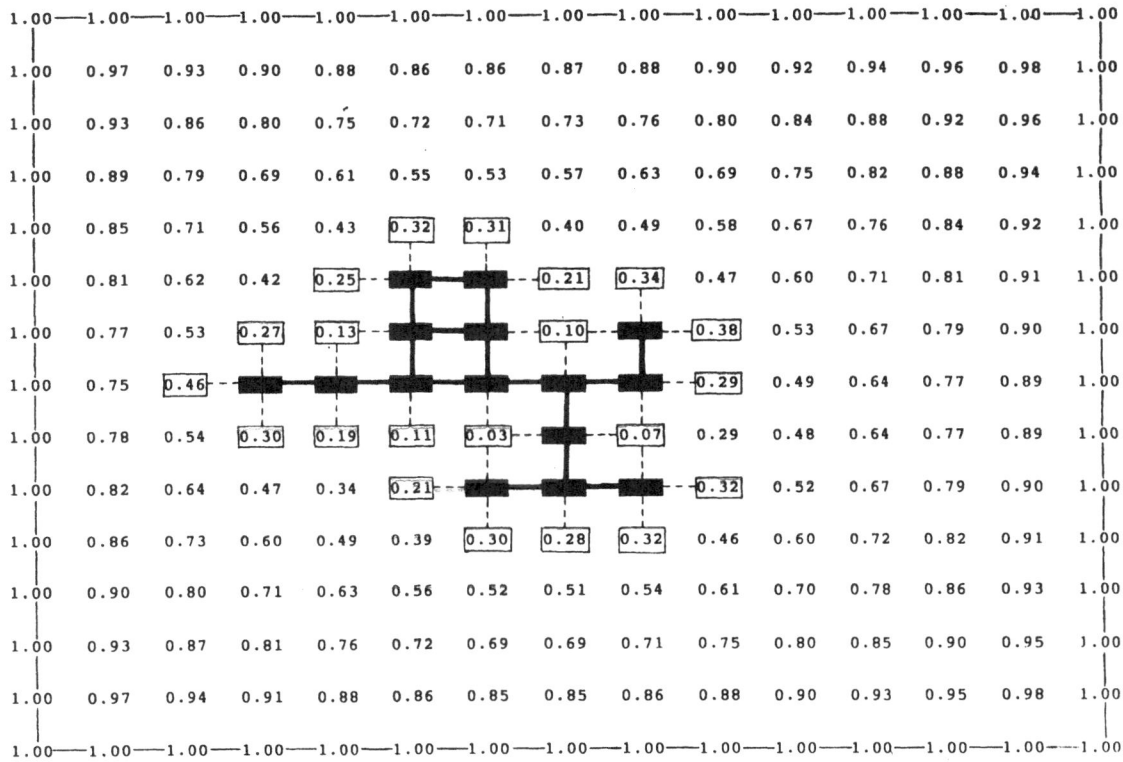

Figure 6.14: The lattice type representation of the potential Φ between a 15–site DBM cluster (at potential $\Phi = 0$) and a square–like electrode that is kept at potential $\Phi = 1$. The site of the DBM cluster, and bonds between them, are marked in black. The sites neighboring the DBM cluster, and the corresponding potential values, are boxed, while their bonds with the cluster are depicted by dashed lines.

Cautiousness is required since the probability law, given by equation (4), does not say that the sites at the tips will be *certainly* occupied (with probability equal to one). The 'law' rather says that a site at the tips has the largest probability, *while* the other sites *may also* get occupied. That is why the branches appear—they start to grow at the sites that, at a certain step of growing the pattern, had smaller probability to become captured into the pattern, but still some of them were captured. Later on, you will see in playing with the program CHILL OUT, how an increase in the value of η favours the tip growing instead of branching of the DBM pattern.

6.6.1   Activity 6: Hands-On: Field Research II

In Figure 14 we present a type of picture you studied in Figure 8, and which is now quite analogous to the picture you can see in Figure 13. Here you are given values of the potential Φ between a 15-site DBM cluster (which is at the potential $\Phi = 0$) and a square outer frame (which is kept at the potential $\Phi = 1$). Of course, the values of Φ have been calculated by the *relaxation method* once the 15–site cluster has been grown. Suppose that we are going to simulate further growing of the DBM cluster and we ask you to estimate the most likely location of the next (the 16-*th*) member of the cluster. Can you do it for us? We are sure you can—just look for the site that is neighboring the DBM cluster and which is at the largest potential compared to all other neighboring sites.

For your own fun, and for the appreciation of the fractal growth process, we urge you to take again your set of pens (different colors!) and to mark: a) the most likely 16-*th* member of the cluster in red, b) the site with the next smallest probability to become the 16-*th* member in orange (if you happen to have the orange color), c) the next site in yellow, and so on up to the least probable site that you can find and mark in dark blue.

At the risk of repetition, we stress that the numbers given in Figure 14, including those that are boxed, represent the Potential Φ of the DBM cluster (the connected little black bricks). A boxed number is proportional to the probability of getting the corresponding site captured into the cluster pattern, but *it is not equal to the probability*, as it should be according to the DBM model (for $\eta = 1$). *Why not? Can you figure out?* Here is our answer: When you add all 21 boxed numbers you get 5.19 (*Check it, please!*), which as a value does not make sense, because this sum has to be equal to 1 reflecting our expectation that one neighboring site will with *certainty* join the cluster in the next moment. *So, what to do?* We simply divide *each* boxed number by 5.19 and we get the proper probabilities that altogether add to one. Thus, in the case of Figure 14, the constant c of equation (4) is equal to $\frac{1}{5.19}$, and, for example, the site that will most likely become the 16-*th* member of the DBM cluster has probability 0.089, whereas the least likely site has more than ten times smaller probability 0.0058 (*Check and locate these numbers!*).

COMPUTER ACTIVITY 3: GROW FRACTALS

[a] You have probably heard that "fractals are everywhere", but we also hope that somebody has told you that every irregular pattern is not fractal. Whoever taught you the latter fact of life, he (or she) was quite right. Fractals can grow in nature and in laboratory, but all

objects that grow and display irregular shapes are not fractals. In this computer activity you can grow fractals *on the screen* using the program CHILL OUT, but you will also be able to grow irregular patterns that are not necessarily fractals. This is a virtue of the DBM model that has been built in the program Chill Out.

[b] Move to the computer program CHILL OUT and click on "Growth". When the small dialog box appears choose the option "New DBM" and you will be asked to select a value for the quantity η. *At this moment we urge you to set $\eta = 1$.* In a minute you will get on the screen a circular border, simulating the outer electrode, and a single point at center representing the cross section of the inner electrode. Now, choose the "Grow" command and the DBM–like structure will start to grow.

You should be patient and wait a while in order to get a picture of the type presented in Figure 15. We hope that you understand the reason for the relatively slow growth. What is your explanation? Is it difficult to answer? No, if you review the stepwise procedure of growing a DBM cluster, you will remind yourself that *each time* a new site appears at the circumference of the cluster, the program makes the computer apply the *relaxation method*, so as to find new potential Φ. Therefore, in the case of Figure 15 and pictures of a similar type, the computer has to go through the relaxation method 1328 times (notice that in the status bar you can read "n:1328" which stands for the number of sites; you can also read the (x, y) position of the last site added to the cluster).

Once you get the picture of the type given in Figure 15, you can go back and compare it with the experimental finding of the Swiss group (see Figure 12). You can see that we may say that, *being a product of a model*, Figure 15 displays the same string–like pattern that was the main feature of Figure 12. The two patterns would get more and more similar if we could continue our "computer growth" to a very large number of sites. Can they get the same? Hardly ever, since both of them are results of random processes. But, the important fact is that one can check (by the methods explained in the second chapter of this book) that the two patterns have almost **the same fractal dimension** (close to $D = 1.7$). Don't think that the mentioned checking is something beyond your capabilities—you can **do it yourself** if you have at your disposal an optical scanner to *digitize* both patterns (one pattern from Figure 12 and the other that you have made using the program Chill Out).

[c] In connection with the remark about the fractal dimensions (of the cluster grown according the DBM model and the pattern produced during the laboratory dielectric discharge), we want to tell you that there is another model that produces, through computer simulations, quite similar fractal patterns. It is the so-called *diffusion limited aggregation model*, or simply the DLA model. If you have not heard of

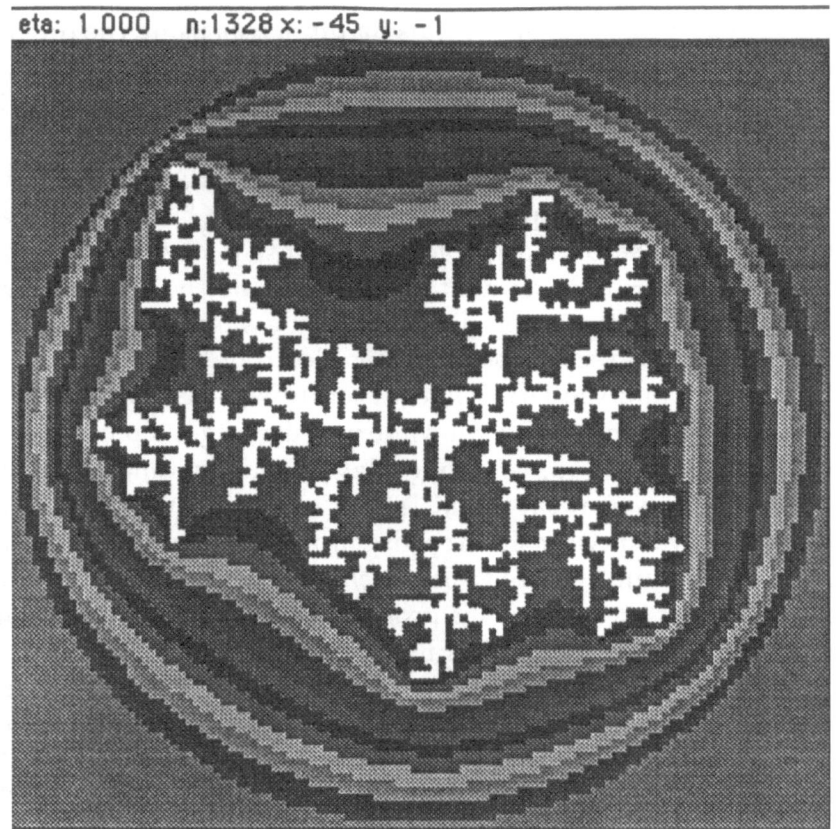

Figure 6.15: The DBM cluster obtained by running the Program CHILL OUT for the choice $\eta = 1$. The cluster consists of 1328 sites, and the last site that joined the cluster is at the location $x = -45$ and $y = -1$ (the center of the "x-y" coordinate system is at the center of the square rim). The shaded bands of constant potential can help you to pinpoint the most probable location of the next (1329-*th*) site that would join the cluster.

it, don't be confused by the fancy three–word name and its abbreviation (DLA). It is a very interesting model that you will be able to grasp and play with (in another chapter of this book). For the time being, you may try to remember that both the DBM model, <u>for $\eta = 1$</u>, and the DLA model afford very similar clusters with the same fractal dimension ($D = 1.7$). Clusters, or fractals, of this type can be found (observed) in nature and grown in a laboratory (including your own school–laboratory).

In the preceding paragraph we have underlined the choice $\eta = 1$ in order to emphasize the fact that for values of η that are different from 1 the DBM model "produces" quite different clusters. We incite you to *"grow clusters on the screen"* for various values of η. For instance, you

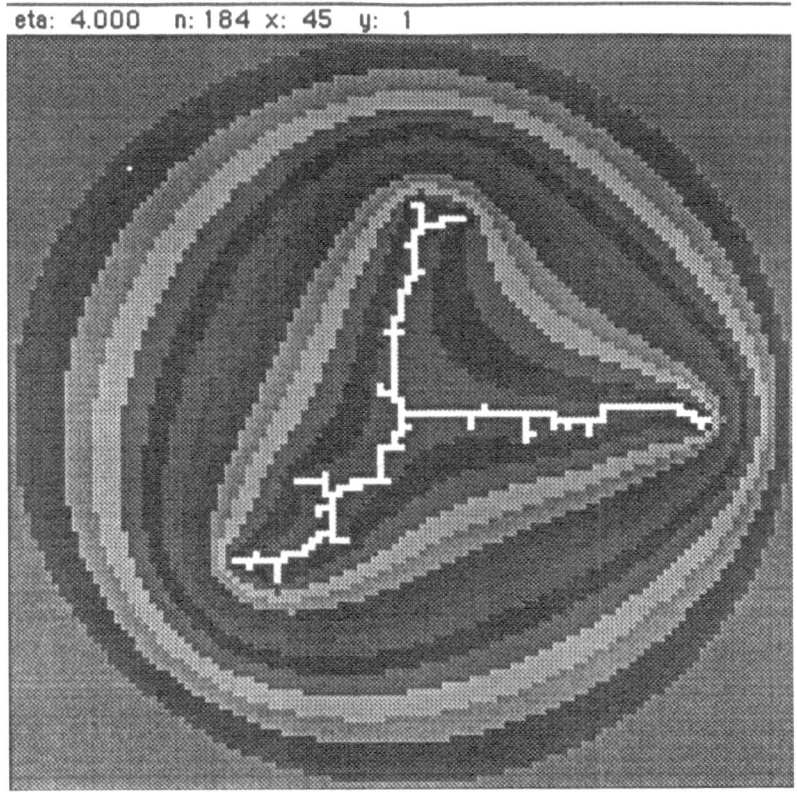

Figure 6.16: A sparse three-branch cluster that has been grown by running the program Chill Out for the choice $\eta = 4$. The cluster consists of 184 sites, and the last (184-*th*) site that joined the cluster is at the location $x = 45$ and $y = 1$, that is, at the tip of the large horizontal branch. You may observe how fast the growing pattern stretches within the available space, in contrast with the choice $\eta = 1$ presented in Figure 15.

may start with $\eta = 0$ and see whether you will "grow" a fractal–like pattern or you will get a hairy ball–like cluster.

For our part, we are offering you our final example in Figure 16—a DBM cluster grown for the choice $\eta = 4$. Why is it so stretched over the screen although it consists of only 184 sites? We may say that it is the work of the "lighting–rod effect" or the "tip effect". What does that mean? Well, you certainly remember that in the DBM model the new sites join a cluster *most likely* at the places where the field is strongest. Indeed, the probability for a neighboring site to join the cluster is given by equation (4) and *when η increases the tips are favoured*. To see that this is true, take any nine numbers, say, 0.1, 0.2, 0.3, 0.4, 0.5, 0.6, 0.7, 0.8, and 0.9 (for the sake of simplicity). Imagine that these numbers represent values of the potential at nine sites that are located next to a growing cluster for the choice $\eta = 1$. Of course, you will rightfully say that the site with the potential 0.9 is at a tip of the cluster. Your statement can be expressed in terms of the probability values following the DBM model prescription (4). Consequently, you should add the nine numbers, getting 4.5, and divide each of them by this total, so that you get probability 0.20 for the tip site to grow. Now, change $\eta = 1$ to $\eta = 2$, and, accordingly, square the nine numbers and divide each of them by their total sum. In this way, you will get a new set of probabilities. What will happen? The new probabilities get smaller, but those that were small decrease faster—compare the smallest 0.022 and the largest 0.20, in the $\eta = 1$ case, and the corresponding values 0.0035 and 0.28, in the case $\eta = 2$. Wouldn't you agree that by changing η from 1 to 2 the tip site has been favoured? Can't we say that for $\eta = 1$ the tip site had nine times greater probability to grow than the 0.1 site, whereas for $\eta = 2$ the tip site has *eighty-one* times greater probability?

We urge you to use your pocket calculator and repeat the probability calculation, with the same initial numbers chosen above, but using $\eta = 4$. Compare your results with the previous two sets of results. Making this comparison you will be able to understand the "tip effect" better. If you have understood it, one of the major goals of this book chapter has been achieved.

6.7   Bibliography

1. A.D. Moore, "Electrostatics", *Scientific American*, March 1972, p. 46.

2. L.B. Loeb, "The Mechanism of Lightning ", *Scientific American*, February 1949, p. 22.

3. E.R. Williams, "The Electrification of Thunderstorms", *Scientific American*, November 1988, p. 48.

4. J.R. Newman, "Laplace", *Scientific American*, June 1954, p. 76.

5. A.L. Kuehner, "Long Lived Soap Bubbles", *Journal of Chemical Education* **35**, 337 (1958).

6. F.J.Almgren Jr. snf J.E. Taylor, "The Geometry of Soap Films and Soap Bubbles", *Scientific American*, July 1976, p. 82.

7. C. Isenberg, "Problem Solving With Soap Films. Part I", *Physics Education* **10**, 452 (1976).

8. C. Isenberg, "Problem Solving With Soap Films. Part II", *Physics Education* **10**, 500 (1976).

9. C. Isenberg, "The Soap Films: An Analogue Computer", *American Scientist"* **64**, 514 (1976).

10. C. Isenberg, "Problem Solving With Soap Films", *Physics Teacher* **15**, 9 (1977).

11. W. Weaver, "Probability", *Scientific American*, October 1950, p. 44.

12. M. Kac, "Probability", *Scientific American*, September 1964, p. 92.

13. B.H. Lavenda, "Brownian Motion", *Scientific American*, February 1985, p. 70.

14. R. Hersh and R.J. Griego, "Brownian Motion and Potential Theory", *Scientific American*, March 1969, p. 66.

15. L.M. Sander, "Fractal Growth", *Scientific American*, January 1987, p. 94.

16. C. Isenberg, *"The Science of Soap Films and Soap Bubbles"* (Dover, New York, 1992).

17. P.G. Doyle and J.L. Snell, *"Random Wlks and Electric Networks"* (The Mathematical Association of America, 1984).

18. R.P. Feynman, R.B. Leighton, and M. Sands, *"The Feynman Lectures on Physics"* (Addison–Wesley, Reading, 1964).

19. E.M. Purcell, *"Electricity and Magnetism"* (Mc Graw–Hill, New York, 1965).

Contents

Chapter 7. Analyzing Rough Surfaces Digitally

Chapter 7

Analyzing Rough Surfaces Digitally

7.1 Introduction

Vast, jagged mountains stretch as far as the eye can see. A spilled cup of coffee slowly conquers the tablecloth. Five minutes after you were supposed to leave for school, your term paper rips as you're trying to separate the pages too fast.

These three different events can produce three starkly different reactions. Viewing mountains, the feeling is profound amazement and utter awe. Watching coffee spread, you might be mildly, self-conscious, because those nearby know you spilled it. Tearing paper leads to hair-wrenching, adrenaline-rushing, voluminous-swearing, bang-your-head-on-the-wall-ing frustration. In this kind of frustration, it's next to impossible to figure out what to do next.

Well, the answer would be to reprint the pages and get your butt out of the house. Nobody sane could suggest that you stop and contemplate life at that moment. But perchance it wouldn't be a bad idea to sit down at one point when you do have the time and figure out what the three situations have in common. A mountain landscape, the advancing edge of a coffee spill, and a paper tear are all rough surfaces, random surfaces, created by complex forces hard to pin down.

Are there some common governing principles that give all these rough surfaces their shape? It turns out that one can predict these shapes to a certain extent, for example how far to one side a paper will tear. What is even more remarkable, these predictions don't require rocket scientists using gym-sized computers. You can make these predictions yourself.

This chapter will show you how.

7.2   Defining the Self-Affine Surface

Before understanding what **makes** a rough surface, we have to understand how to **describe** a rough surface. Anybody can see the difference between tabletop and the profile of a mountain: one is smooth, the other is rough. But can you see the difference between two mountains? Is one mountain rougher than another? To answer this, we have to understand the **geometry** of a mountain. It is neither a circle nor a square, it is something more complicated. It is a mountain. In this section we show that there are some very simple geometrical objects, called **self-affine objects**, which may look like a mountain, and which are extremely helpful in answering the question: How rough is a mountain?

7.2.1   The Reading Glass

Look at the circle in Figure 1a. Now take a reading glass and view the circle through it. You see a bigger circle, like the one on the right. What did the reading glass do to the image of the circle? It increased its size by enlarging it both vertically and horizontally. You may observe that the circle remained a circle, only it became bigger. What the reading glass did was, say, double the vertical and horizontal sizes. If originally the height was 1 cm, and the width 1 cm as well, then the big circle would be 2 cm in both directions.

But what if we have a peculiar reading glass, which changes the vertical and horizontal dimensions by different amounts? For example, while it doubles the vertical direction, it increases the horizontal size by 4 times. Let's see what's going to happen with our circle in this case. If we take our original circle of size 1 cm horizontal and 1 cm vertical, it becomes a 4 cm by 2 cm ellipse, like you see in Figure 1b.

It is not only the circle which is different under this glass. If we view a tall girl, or a family car, under our peculiar glass, they will resemble a short girl or a limousine, respectively, as in Figure 1c. The girl still looks like a girl, but she is not the same any longer: she seems stubbier now. Similarly, our family car became a limousine. However if we view the profile of a mountain it looks the same after magnifying with our glass. As you see, some objects change under our peculiar glass while others, like the mountain, remains invariant and looks the same.

Could you imagine how other, real-life objects would look under this glass? Do you think that there are more objects which look just the same under the glass as without it? You are about to construct a simple model of an object which **looks the same** under this peculiar reading glass as it does without the reading glass. This kind of object is known as a **self-affine object**.

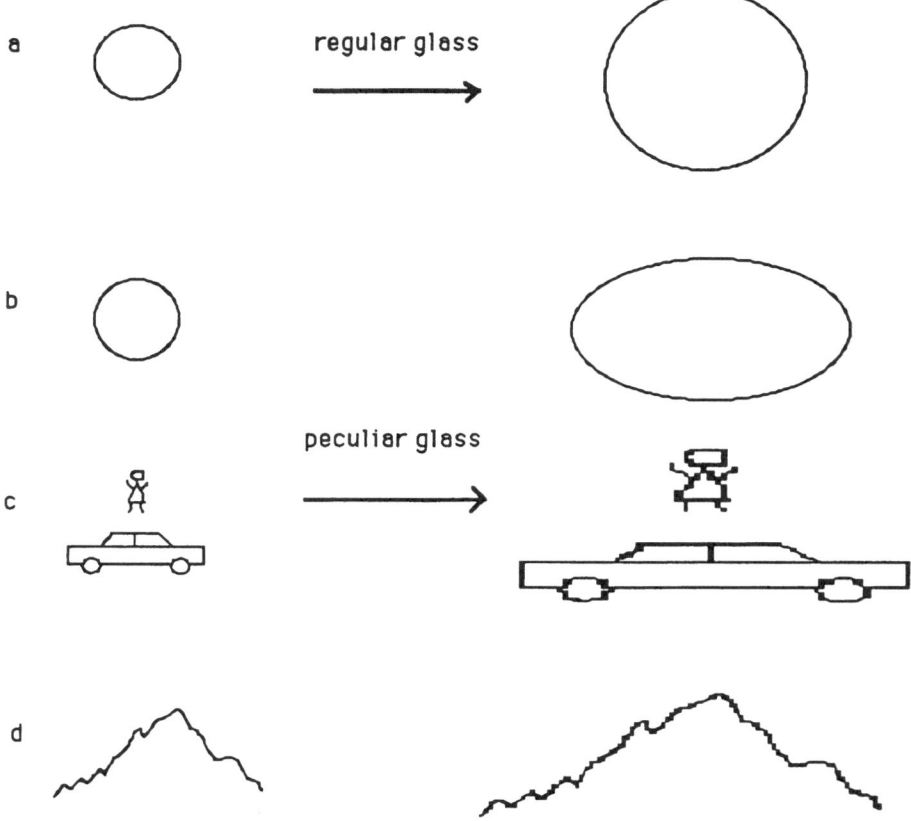

Figure 7.1: Regular and "peculiar" magnifications.

7.2.2 Constructing the Self-affine Object

Start constructing a sample self-affine object by drawing a diagonal
from the bottom left to the top right of a rectangle of length 16 and
height 4 in arbitrary units (see Figure 2). The two ends of this diagonal
are fixed, and, in the next step, this line is divided into *four* segments:
one up, one down, and two more up. These last two segments need
to gain the same height as the original diagonal did, so, since they
only have half the domain in which to gain the height, they need to be
twice as steep. We make the other two segments slope up and down by
the same amount as these. This factor, the factor of the slope being
twice as big as before, will become very important in making the object
self-affine.

We repeat this step again and again, dividing every line segment into
four segments twice as steep. The rule is this: Every "up" becomes
an "up-down-up-up", and every "down" becomes a "down-down-up-
down". The process can be repeated infinitely. However, after several

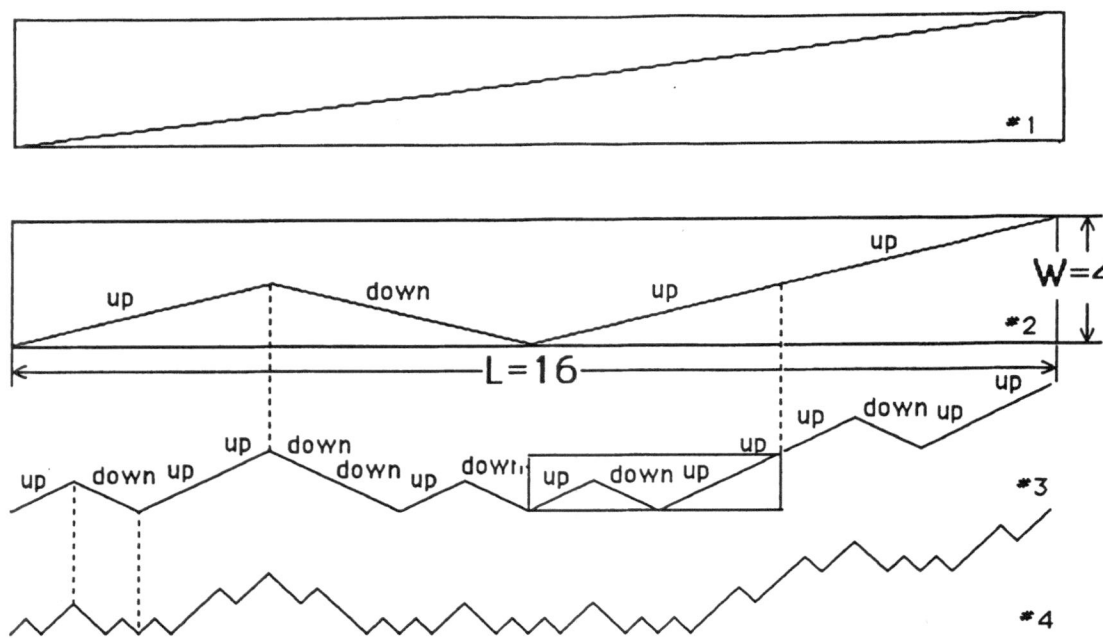

Figure 7.2: Creating a self-affine object.

iterations the details of the surface get too small to see. Let's stop the iteration process in the third step and see what happened if we look at the resulting object using our peculiar glass. If we take a portion of the profile of length 4 (see small box in Figure 2) and enlarge its length by a factor of 4 and its height by a factor of 2, we see that we recover the same object we had in the previous iteration (see Figure 2). The same magnification can be done using different parts of the profile in different stages of the iteration process and we are going to get the same result. Thus, we have generated an object that looks the same under a glass that magnifies it with different factors in each direction. This property of being invariant under this peculiar glass is what define a **self-affine object**. If the construction process were carried out for an infinite number of steps, the result would an example of an **anisotropic fractal**. (The word "anisotropic" means "different in different directions")

Maybe you can think up another rule to generate a different self-affine object for which you need *different* magnification factors in the glass. This different relative magnification gives us a way to describe such objects. Different self-affine objects may need different magnification factors if we want to see the same picture after looking at them through the peculiar glass. In a more "scientific" formulation this means that differents self-affine objects remains invariant under transformations of different magnification factors.

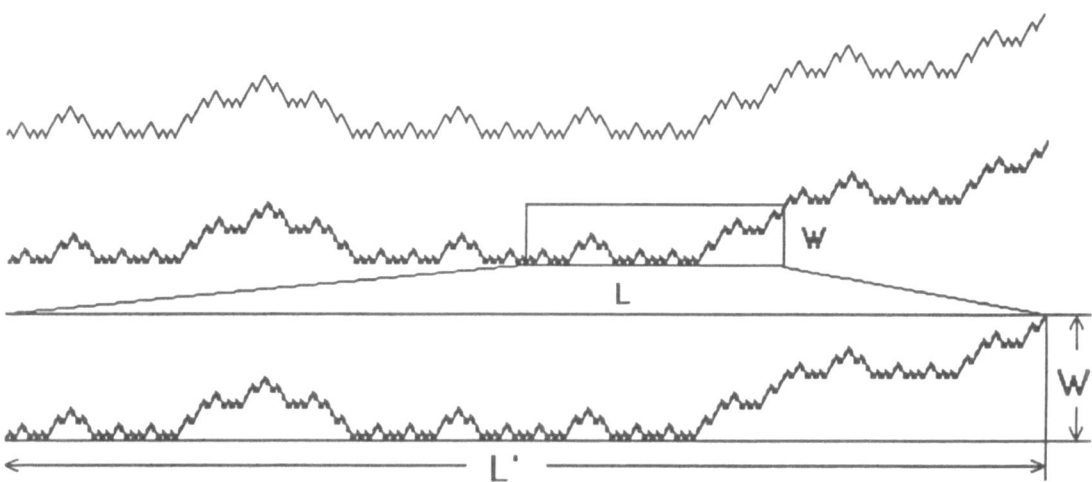

Figure 7.3: Finding the self-affinity factor.

From Chapter 4, we remember a way to characterize a DNA landscape by using what we called the **roughness exponent** α. Let us apply that definition here. Instead of using the profile of Figure 4.5 of Chapter 4, let's use the one in Figure 3 of this chapter that corresponds to the third iteration that we have just generated. Following the directions given in Activity 17 of the DNA Chapter, we can measure the width W of the profile as a function of l, the length of the box we use to calculate the width. You can do this for the following values of $l = 1, 4, 16$, using Figure 2, and writing down the results in the following table.

Width versus length	
l	W
1	
4	
16	

First of all, you see that the width W increases with the window size l. If you use a log-log paper to *plot W* versus *log l* you can see that the points lies on a straight line. You probably remember from previous units that every time that a log-log plot gives a straight line, this means that there is power law relation between both quantities. This power law relation between the width W and the length l is a mathematical way to express what we have learned from our more friendly peculiar reading glass, as follows

$$W(l) = A \ l^{\alpha}, \tag{7.1}$$

where α is what we call the **roughness exponent**, the quantity that characterizes a given self-affine object. On the other hand, the multiplicative factor A is a constant that we don't care about because it only depends of the units we use to measure the length and the width (for example, centimeters or meters).

To relate this equation to the result of the log-log plot we take the logarithm in both sides of the above equation, and using some famous properties of the log function we get

$$\log(W) = \log(A \ l^{\alpha}) = \log(A) + \log(l^{\alpha}) = \alpha \ \log(l) + \log(A). \tag{7.2}$$

Thus, we see that by plotting W versus l in a log-log paper we can get the roughness exponent α from the slope of such a plot. We also see that we don't have to care about the factor A since it doesn't affect the value of the slope (i.e., the value of α).

You will find that the slope of our plot is 0.5. Thus we get our first self-affine object and characterize it with a number: the roughness exponent $\alpha = 0.5$.

As an extra activity, you can calculate the width as a function of l using the next (fourth) iteration of the self-affine profile (Figure 2) or you can write a short code for a computer to generate the profile up to a certain iteration stage like in Figure 3, and calculate the exponent there.

Now that we have learned how to describe a self-affine profile, we can go back to our original question. How rough is a mountain? Well, we can say that we have a new quantity, the roughness exponent, that can answer this question: the rougher the mountain is, the bigger is the value of α.

7.3   Activity: The Paper Tear Experiment

A self-affine surface doesn't have to be invented by an iterative procedure with a definite formula, such as the one we just used. Many natural, random **rough surfaces** exhibit the property of self-affinity as well. For example, we have learned that the concept of roughness can be applied to random walks and landscapes in DNA and literature as well. But now we are interested in self-affine **surfaces** like the profile of a mountain. We will see that we don't need to "create" a mountain to have such an object. We will perform a very simple experiment to create a rough surface that can be understood with the same concepts we used to describe the mountain. This is the paper tear-experiment.

7.3.1 The experiment

A paper tear is created by pulling at the two ends of a sheet of paper. A good material to use for producing a paper tear is a paper towel without perforation. Since it will have to be scanned into a computer, dark colors are preferable. If you cannot find a paper towel dark enough, it can be dyed after making the tear.

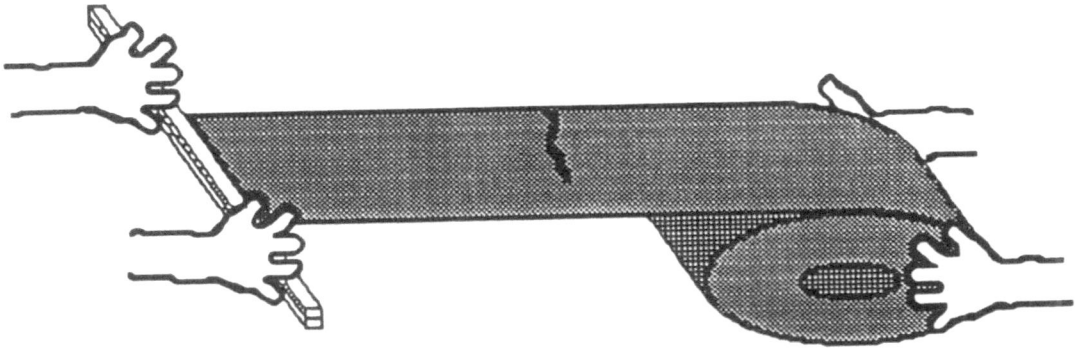

Figure 7.4: Tear analysis experiment.

You will need to obtain a roll of unperforated paper towels, used in many dispensing machines in public bathrooms. Ask the janitor or custodian for such a roll. Then, find a partner, unroll a sheet about 50 cm long, and make a small notch in the middle of the edge (Figure 4). One partner holds the end of the sheet between two yardsticks or pointers while another holds the bulk of the roll. If the two of you, simultaneously and evenly, pull the two ends apart, the sheet will crack at the little tear, making an uneven edge which is the subject of all the subsequent analysis. Sometimes, when you try to rip the paper in this manner, it will tear right next to where you're holding it. In that case, unroll more towel, make bigger notch a little, and try again.

Now look at the profile of the tear; it is similar to the mountain! It is rough! How do we characterize it? Remember: by measuring the roughness exponent α.

7.3.2 Measuring the roughness exponent with the computer

To measure the roughness exponent of this paper tear, you will have to measure how far away from the straight line across the middle the paper tear gets after going a certain distance. Of course, these measurements can be made by hand with a ruler (as we did in the previous section), and it probably wouldn't hurt to make a couple of rough ones yourself. But if you do this you'll soon discover that this thing needs a lot of measuring, and it would be much easier to have the measurements made by a computer.

7.3.3 Scanning the paper tear

This particular part of the experiment was developed using a Macintosh with a scanner, so you'll need a Mac and certain applications for it. What you need to do is scan the torn sheet of paper using an application such as AppleScan and an Apple scanner itself. This way the computer will have a digital image that it can deal with.

Let's turn now to the computer and learn how to obtain a digital picture of the profile of the tear. You need to follow the following steps:

1. Before running the application, create a folder in which save all the outputs files. Choose **File** at the top of the screen and click on **New Folder** on the menu. Put a name to your folder.

2. After turning on the Apple Scanner, restart the computer, otherwise the Mac won't recognize the scanner. To restart the computer choose **Special** at the top of the screen and click on **Restart**.

3. Put the paper tear on the scanner screen and cover it with a black sheet of paper to increase the contrast between the paper and the background.

4. Open the folder **Applications** and run the program **AppleScan** . You can accomplish this by double clicking on the respective icons. You will see a small window at the upper left corner that we'll use to preview the image of the paper tear. Another window named **Untitled 1** will appear on the screen. This is the document where we are going to save the image.

5. Choose **Control Panel** in the **Scan** menu in order to set a the following parameters:

   • **Resolution** must be at **75 dots per inch**
   • **Threshold** must be around a value of **10**
   • **Line Art** must be selected

Before scanning, we recommend that you preview the image to check that all the parameters have the correct values. Click on **Preview**. If your picture doesn't fill the whole screen you should reduce the frame to scan only the profile of the paper tear. To this end you can click on one of the four corners of the frame and reduce/enlarge it.

6. When you think that your picture is ready, click on **Scan** (we are still in the **Control Menu Panel**) to obtain the final scanning. The image will appear now in the document **Untitled 1**, so before scanning, be sure that there is a document with that name on the screen (that you didn't close it by mistake). You can zoom in the image by peaking **Zoom in** at the menu **Goodies**. If you are not satisfied with the image you can always go back to point 5 above, make the appropriate changes, and restart the process again. For example, a common problem is that you see many black dots on the paper. This is an indication that your **Threshold** is too low. So, you may want to change it (to 11 or 12, for example) and redo the scanning.

7. Save the image picture now. Choose the **Tools** menu and click on the dashed square at the middle of the second row from the top. Select the part of the picture in the document **Untitled 1** you want to save. Choose **Save as** from the **File** menu and save the file in the folder that you created. Make sure you save the image as a file of MacPaint.

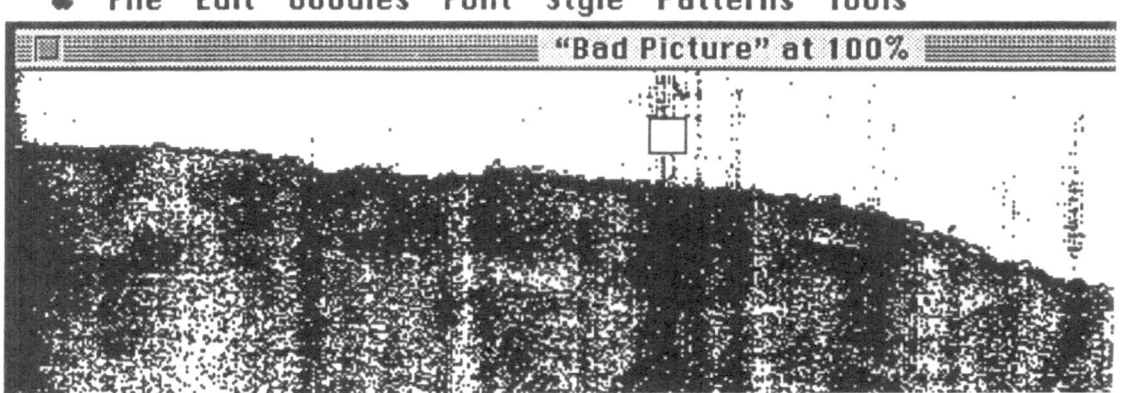

Figure 7.5: Erasing the dots not connected with the interface.

8. Because it is impossible to produce a perfect image with the scanner, you need to clean up the image manually. This is accom-

plished by opening the saved MacPaint file (clicking twice on it).
Choose the eraser from the **Tools** menu (second in the bottom
row) and erase the dots that appear above the edge of the paper
and are not connected with the interface (see Figure 5). If necessary, you can zoom in to do this. Also, make sure that the edge
is solid, that there are no holes in it. You also need to rotate the
image so that the black surface is on the bottom, as it is in the
picture. This is done by clicking on **Invert** in the **Edit** menu.
Finally, you have to save the resulting image using **Save as**, as
before.

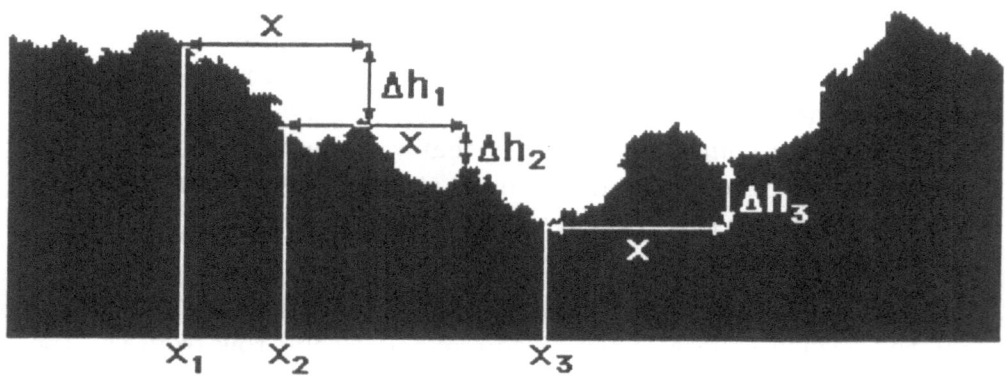

Figure 7.6: Analyzing the surface.

7.3.4   The Rough Surface Analyzer Program

Now the image is ready to be analyzed. The Rough Surface Analyzer
application was written just for this; open it first and then open your
MacPaint file from it. The program will make the same measurements
that it would have taken you so long to do by hand. More specifically,
it will take a set of calipers set for a specific distance x, and move the
beginning point sequentially from x_1 to x_2 to x_3 and so on (see Figure
6). Then it will calculate the width (as we did before by hand) for each
starting point, and average all these values to obtain the average width
$\Delta h(x)$ for each value x. Finally, it will plot *log* Δh versus *log* x and fits
a straight line for the data points (Figure 7). It will do this both for
the paper tear-surface that you give it and for a random walk surface
that it will generate (see next section).

In this graph, the x and the Δh are equivalent to l and w for the
self-affine figures, respectively. The slope of this line, determines the
roughness exponent α of our paper tear. Chances are the slope will not
hit the actual value right on; it can sway between 0.5 and 0.7, or even
wider. If it's not, what should you do to get a closer outcome? Should
you average the results of several paper tears to get a better result?

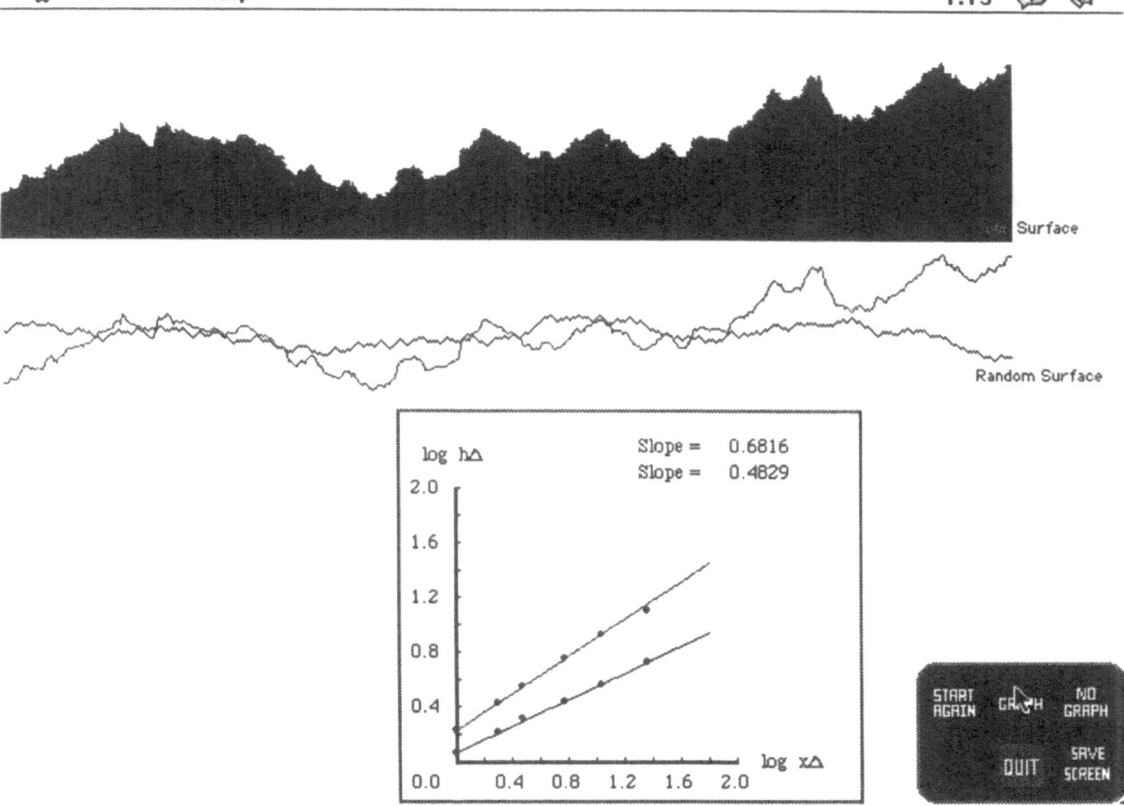

Figure 7.7: Analyzer results.

If this is done many enough times, you will see that the average α is some-where very close to 2/3. This is the roughness exponent of the paper tear.

7.4 The Paper Tear: Understanding Experiments with Models

A sheet of paper may look pretty uniform from the outside, but inside it's made up of distinct fibers that have connections of different bonding energies between them (Figure 8). In order for the paper to tear, these connections need to be broken along some profile. Thus the paper tear follows a route along which the bonds between the fibers are easiest to break or have the minimum energy.

This effect can be simulated by a triangle of heads-up pennies (Figure 9). The **year** on each penny corresponds to the **energy** of a bond at that point. The penny on top corresponds to the point where the tear

Figure 7.8: Fibers in a sheet of paper.

starts, and the route from the top penny to the bottom row corresponds to the route of the tear.

Up until several years ago, people knew only one algorithm for how a paper tear might choose its route. This algorithm is the so-called **random walk model**. At every step, the tear chooses the penny with the smaller year (corresponding to the bond with smaller energy) and goes that way. The years of the pennies are uncorrelated to each other, thus each step is completely unrelated to the previous one. Presuming that the years of the pennies are randomly distributed throughout the triangle, the way that the path goes at each step must be random. It is a random walk. That is the prediction of the random walk model.

In the model, the fibers, represented by pennies, are arranged in a neat, orderly pattern, something that doesn't occur in a real sheet of paper. Moreover, the tear can only go down without turning up, and its options at every point consist of **locally** choosing either the penny below right or below left. Certainly these limitations confine the accuracy of the model to a certain extent. Any model's creator, though, must sacrifice a certain degree of accuracy for increased simplicity. This is the essence of making a model. What distinguishes a good model from a bad one is how well it can simplify the problem while still predicting results near to observed values.

Let us now look at the results of a tear modeled on the random walk.

7.4.1   The Random Walk Model

To visualize a random walk, let us pretend that a drunk man is playing with the elevator in the World Trade Center. There are two buttons

Figure 7.9: The "random walk" route.

in his elevator: "Up one floor" and "Down one floor'. When he is on a certain floor, he presses one of the buttons. When he gets to the next floor, he presses a button again completely un-related to the one he pressed the last time (Figure 10). Let's try to find out if he gets anywhere eventually.

Our drunk man has been playing with the elevator every day for the past year, and we keep an eye on him in order to plot his trajectory. We gather information to graph the average distance he move away from the floor where he started versus the number of "steps" (up or down a floor) that it take him to where he is now. If we keep track of him for a very long time, we come to the conclusion that the square of the distance is proportional to the number of steps. It's not all that hard to prove that this is true on the average for a random walks; try the method of induction if you want to do it or review Chapter 1.

In Figure 11 we plot the log of the average distance as a function of the log of the numbers of steps of the drunk man. They play the role of

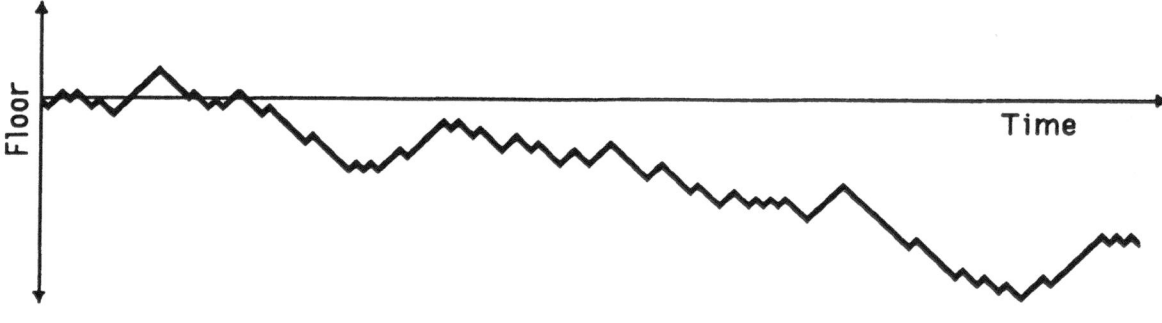

Figure 7.10: Random walk of the elevator.

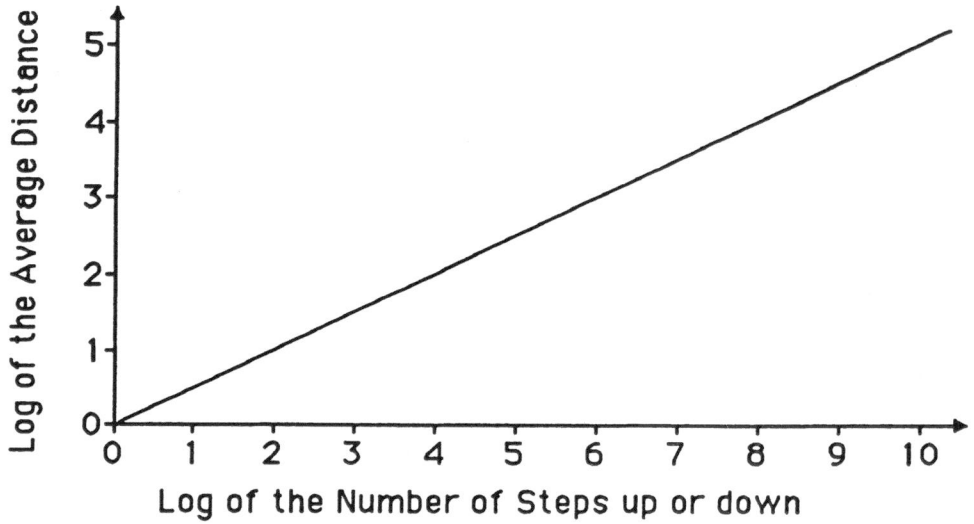

Figure 7.11: Progression of the random walk.

the width W and the length l of a self-affine profile, respectively. The slope of $1/2$ of the straight line in the plot represents the roughness exponent. **Thus a random walk is a self-affine surface with an α of 0.5!**. In Figure 12 we see that, especially for the left half, an enlarged portion of the surface looks almost identical to the surface itself. Of course, since the surface is a **random** walk, this resemblance is never absolute, and it's never absolute for a paper tear either. But self-affinity does not require the enlarged image to be identical. This property is only exact for deterministic self-affine objects like the one we define in Figure 2. Our random walk profile is self-affine if the width is a certain constant power of the horizontal length.

This power, the roughness exponent, is a good litmus test to see if the random walk model fits the paper tear. That is exactly what we

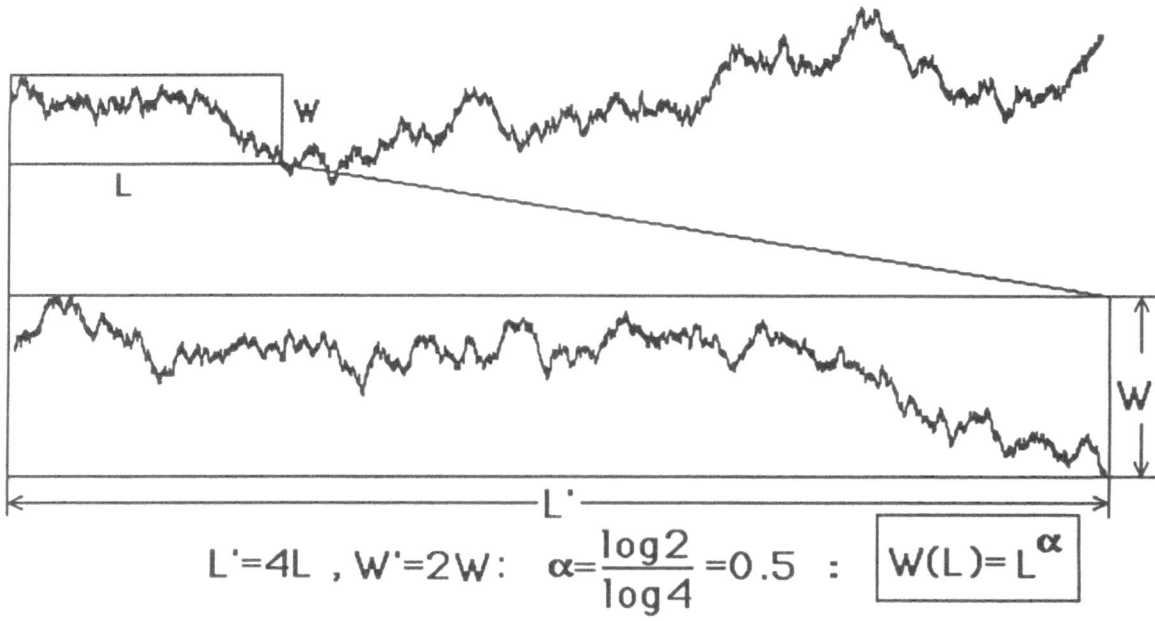

$$L'=4L \ , \ W'=2W: \quad \alpha=\frac{\log 2}{\log 4}=0.5 \ : \ \boxed{W(L)=L^{\alpha}}$$

Figure 7.12: Self-affinity in a random walk.

can check right now. The profile of the random walk can be plotted next to the paper tear along with the width of the profile as a function of the length in the paper tear experiment with the **Rough Surface Analyser** application (see Figure 7). The slope for the random walk line should have been 1/2 (0.4829 in Figure 7), right ? A good estimation for the roughness exponent of the paper tear should give 2/3 (0.6816 in Figure 7). What does this mean? This means that the random walk model doesn't fit the paper tear. Cáspita! We need more theory!

7.4.2 The Minimal Energy Route Model

In 1991, a new theory was developed that explains the paper tear much better than the random walk. This alternative, more complicated, model is called the **minimal route**. In contrast to the random walk model, where the tear chooses the minimum energy route only for the next step, in the new model, the tear follows a path that requires the least **total** or global energy in order to break the bonds between the strands of paper. In this new model, the tear must check the energy of the entire path and choose it if its energy is the smallest of all possible paths.

Figure 7.13: The two models.

A penny triangle of six rows or so would be a good size to explain the model (Figure 13). The year on the penny determines its energy. We want to find the minimal total energy route from the top penny to the bottom row. That is, we want to find that route for which the years on the pennies add up to the smallest number. Of course, we can do this by adding up the years for all possible paths, but that would take a very long time to do. There is a much faster algorithm for doing this.

What we do is look at the second row from the bottom. If the minimal energy route were to come to the leftmost penny in that row, where would it go next? In this example, it has two options: the '76 penny and the '73. It would go to the '73. Like this we go across the row and see which way the route would go if it were to come to that point. In some cases, for example the last penny in the row, the choices are equal. In such a situation we choose randomly, let's say the penny to the right.

When we are done with the second row from the bottom, we go to the third row from the bottom. If the route were to come to the left most penny here, which route would it take from that point? We add up the years for the two pennies in each of the possible routes, and choose the smaller sum. In this example, the route would go to the '61 and '73 rather than to the '85 and '73. In a similar manner, this we keep going across and up until we come to the topmost penny. Now we know the minimal energy route.

As you can see, finding the minimal energy route is not exactly a piece of cake. It would take a person playing with pennies ten or fifteen minutes to calculate it for a triangle with six pennies to a side. A sheet of paper is around one thousand fibers across, corresponding to a triangle with 1000 pennies to a side. If we checked all possible routes there are 2^{1000}, or 10^{300} different ones from the top to the bottom. If it takes a computer program a millionth of a second to calculate each route, then it would still take the computer 10^{286} years to check all the routes. By comparison, the universe is considered to be "only" 10^{10} years old.

With a simplified algorithm such as the one described above, the number of routes the computer needs to check for a triangle of n pennies is around $n^3/6$ instead of 2^n. Thus for a 1000-fiber sheet, the computer would need to check only 167 million routes. The fastest supercomputers in the world, the ones that can carry out a million operations a second, would need a little less than three minutes to do what a simple sheet of paper accomplishes in a fraction of a second: find the minimal energy route for a tear. That is one smart sheet of paper you are dealing with!

7.4.3   How the Paper Finds the Minimal Energy Route

Let's assume that the paper is composed of n fibers each of which that can resist a very small threshold force f_i. If the force applied to a fiber exceeds this threshold, it breaks. Here these forces play the same role as the energies in the previous section. When we pull the paper at the ends, the external force F is transmitted through the fibers in the paper. For simplicity's sake, let us suppose right now that the strands never intersect, but just run parallel to each other like the hairs in a braid or the fibers in a rope (Figure 14).

To understand how the paper breaks, you can make a simple experiment with a thread supporting a saucepan to which you add water or sand. How does it happen that a rope made of many such thread can hold a much larger weight ? Because the force is distributed between all threads

$$F = f_1 + f_2 + \cdots + f_n , \qquad (7.3)$$

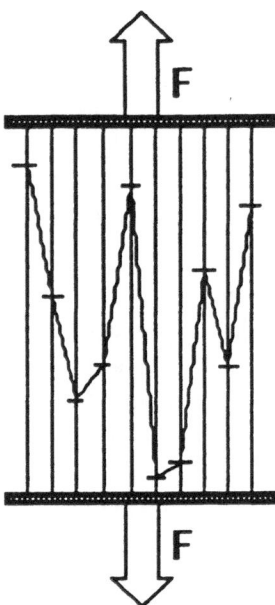

Figure 7.14: The breaking of parallel fibers.

where each of the small forces f_i is smaller than the threshold t_i for the corresponding fiber. When the force F exceeds $t_1 + t_2 + \cdots + t_n$ the rope (paper) breaks.

The fibers in a sheet of paper intersect and are connected. They can be represented more accurately by a lattice (Figure 15). This lattice, which is exactly the same as the triangle we saw before (only turned sideways) breaks along a path from the notch in the middle of the left side to the right edge. Let us suppose that the lattice is very rigid, so each fiber along this path must be broken for the lattice to break. Thus the breaking force must equal or exceed the maximum force that the total of all the fibers along a path can endure.

As the forces pulling on the two ends increase, more and more tension is applied to each fiber. When the total force crosses the threshold of endurance of the weakest path, all the fibers along that path break. This path, because it was the first one to break, is the minimal energy route. Rather than figuring out all the possibilities, the sheet of paper finds the route by testing which one "gives" the easiest.

This lattice model that provides a simple explanation for how a tear forms in a sheet of paper also describes the cracking of a brick wall. When a wall in a brick building cracks during an earthquake, all the cement "fibers" between the bricks break at the same time, and the building "splits at the seam" down the path of minimal energy.

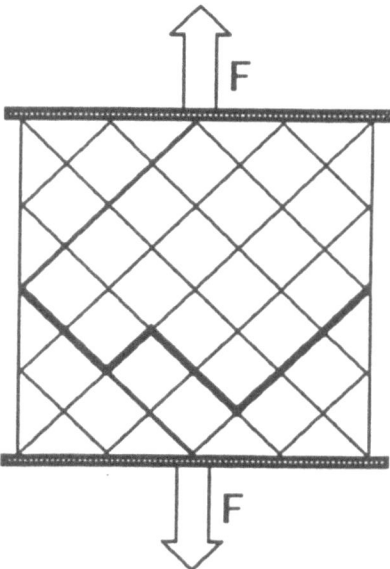

Figure 7.15: The breaking of intersecting fibers.

7.5 Testing the Minimal Energy Model

Now that we have minimal energy model, we still have to find out its self-affinity factor to see whether or not it fits the paper tear. While it is possible to do this theoretically, the calculations turn out to be unbelievably complicated, and still at times shaky. It would be much easier to test the model numerically, try it out in a theoretical situation, do the measurements, and calculate the self-affine exponent. In fact, the result $\alpha = 2/3$ was obtained by an student who used computer simulations and only later was this result obtained analyticaly. The **Dirpol1** application on the Mac does just this.

Open the program and go to the **Change parameters** option in the **Control** menu. The program will ask you to enter the **length parameter**. This length parameter is nothing more than the length of a side of our good old penny triangle, only here the program will represent the triangle as a lattice. 10 will be a good size for now. The other parameter, μ (mu), in the form of a fraction, controls how evenly the strengths of the bonds are distributed. The larger the m, the more even the distribution. Don't worry too much about this parameter; in fact, nobody knows for sure just how evenly the different strengths of the fibers are distributed in a sheet of paper. A μ anywhere from about 5 to 10 will do.

The best way to get into the program is to play the game that comes with it. Choose **Game** from the **Speed** menu, then go to **Start** in the **Control** menu. In this game, you will have a few seconds to guess the minimal energy route in the lattice that the program gives you, and click on the bonds that the route will go through. Experts at this game know that the route will almost always pass through the weakest (red and yellow) bonds. With this kind of insight, you should become an ace at this game after a little practice.

After your name obliterates the record charts, you can get to some of the more serious aspects of the application: measuring the self-affinity factor of this model. Choose a larger lattice size (20 would be good), and switch the speed from **Game** to **Lightning**.

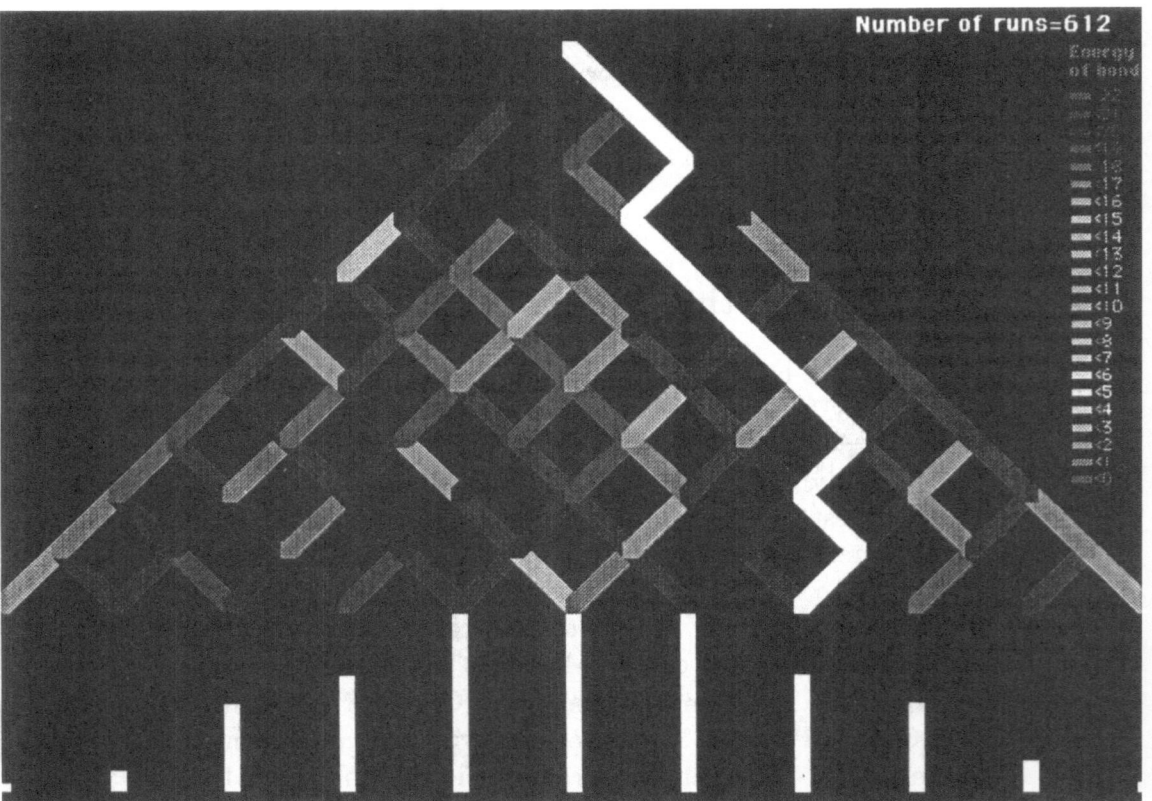

Figure 7.16: Minimal energy route histogram.

Now the program just shows you the minimal energy route of a different lattice every second or so, and builds up a histogram on the bottom of the screen of where the route ends up (Figure 16). You need lots of data to do the analysis, so switch the speed to **Invisibly fast** and let the histogram grow until the number of runs gets to about 500.

Even though you did not see it, **Dirpol1** has been doing some serious analysis all this time. It was figuring out the same things that Rough

Surface Analyzer figured out for the paper tear: how far away from the middle line the path gets with every specific number of steps. Choose **Average** from the **Control** menu to see these results plotted on a log-log scale. If you click on two endpoints to get the slope of the line, you will see that it is indeed somewhere right around 2/3. Thus the minimal energy route does fit the paper tear!

7.6 Conclusion

And so we have seen the development and testing of two models of a rough surface, predictions that did or did not pan out. For other rough surfaces, such as the creeping edge of spilled coffee, mountain landscapes, surfaces of bacterial cultures, models like this have to be constructed, and predictions made. For all of these rough surfaces, a lot still remains to be done in the exciting process of exploring hypotheses. For each one of them, including the paper tear, nobody has all the answers.

Currently, the minimal energy model seems to fit the paper tear extremely well. That does not mean, however, that it is the final word to be said on this issue. Perhaps when you analyze the paper tear more closely, you'll see a discrepancy between it and this model, and think of a new, better model. The future of this science is up to YOU. Don't give up!